SHOW ME
THE HONEY

SHOW ME THE HONEY

ADVENTURES OF
AN ACCIDENTAL APIARIST

DAVE DOROGHY

TOUCHWOOD

Edited by Claire Philipson
Copy edited by Meg Yamamoto
Cover design by Tree Abraham
Interior design by Sydney Barnes

LIBRARY AND ARCHIVES CANADA CATALOGUING IN PUBLICATION

Title: Show me the honey : adventures of an accidental apiarist / Dave Doroghy ; foreword by Rick Hansen.
Names: Doroghy, Dave, author. | Hansen, Rick, 1957- writer of foreword.
Identifiers: Canadiana (print) 20190193468 | Canadiana (ebook) 20190193492 | ISBN 9781771513227 (hardcover) | ISBN 9781771513234 (HTML)
Subjects: LCSH: Bee culture—Anecdotes. | LCSH: Beekeepers—Canada—Anecdotes. | LCSH: Doroghy, Dave.
Classification: LCC SF523.3 .D67 2020 | DDC 638/.1—dc23

TouchWood Editions gratefully acknowledges that the land on which we live and work is within the traditional territories of the Lkwungen (Esquimalt and Songhees), Malahat, Pacheedaht, Scia'new, T'Sou-ke and W̲SÁNEĆ (Pauquachin, Tsartlip, Tsawout, Tseycum) peoples.

We acknowledge the financial support of the Government of Canada through the Canada Book Fund and the Canada Council for the Arts, and of the Province of British Columbia through the British Columbia Arts Council and the Book Publishing Tax Credit.

Canadä Canada Council for the Arts Conseil des Arts du Canada BRITISH COLUMBIA BRITISH COLUMBIA ARTS COUNCIL An agency of the Province of British Columbia

This book was produced using FSC®-certified, acid-free papers, processed chlorine free, and printed with vegetable-based inks.

Printed in Canada

24 23 22 21 20   1 2 3 4 5

To my mother, Susan, who always taught my sister and me
that the more things in life you are interested in,
the more interesting your life becomes.

CONTENTS

FOREWORD

I've known Dave for over 30 years, and he often jokes that I introduce him as "Dave, the honey guy." In recent years, I've indeed enjoyed jars of honey from Dave's hives, but you might be surprised to hear that our connection started with a global burger chain. Long before Dave became the honey guy he was my "McDonald's guy."

A series of fortuitous events led Dave to join my Man In Motion World Tour team. I had been on the road and all around the world for over a year, but the Tour was far from a fundraising success. In fact, I was on the verge of losing hope. At this time, Dave was a junior ad man with Palmer Jarvis advertising agency, and for over a year had been on the McDonald's account doing advance promotion work of the Tour from Vancouver. This involved developing promotional materials, fundraising manuals, and even sewing McDonald's patches onto the track outfits that Nike had donated.

It was nothing short of providence that George Cohon, chairman of McDonald's Canada, happened to spot me on NBC's *Today* Show with Bryant Gumbel talking about my journey around the world in my wheelchair. What also caught his eye that day was the logo on my shirt, one of the ones Dave had diligently sewn on to garner exposure for the restaurant.

George was passionate about worthy causes, and I was lucky that my dream to find a cure for spinal cord injury and to raise awareness of the potential of people with disabilities was one he wanted to get behind. George contacted Ron Marcoux, McDonald's Western Canada vice president, to bolster McDonald's existing sponsorship of the Tour. Dave possessed an entrepreneurial spirit and energetic attitude, which made it a no-brainer for his boss, George Jarvis, to ask Dave to pack his bags for the year to join me on the last leg of the Tour from the East to West coast of Canada.

Not only was Dave tasked with being my advance man, he was also given the goal of helping McDonald's raise a quarter of a million dollars for the Tour on the homestretch. This was no simple feat, but Dave rose to the challenge. Over the course of nine months, Dave and I forged a friendship along the Trans-Canada Highway.

I marvelled at how organized each McDonald's restaurant was when I arrived in town. Each one welcomed me with a banner that said "We're with you Rick." McDonald's, I thought, must have had to produce hundreds of those welcome banners. I was surprised to find out later that there had only been one banner! After

I left the restaurant in each town, Dave and the local restaurant manager would set up a ladder and climb up on the roof to unravel the signage. The banner would be rolled up and shipped on a Greyhound bus into the next town, ready to go up on the next McDonald's roof before I arrived.

Dave was one of our most energetic scout bees, and this is just one of many stories that I could tell you about his relentless energy and professionalism. I will forever be grateful for Dave's role in the success of the Tour, but even more so for his friendship and support in the years that followed. Dave has always possessed a natural curiosity and ambitious spirit that allows him to jump into sticky adventures with ease. It came as no surprise to me that Dave had begun chasing new dreams as an apiarist. As I read his memoir, I was struck by the parallels between building a thriving bee colony and the fine margins between life's victories and failures.

It can be easy to look back on memories of the Tour and only remember its triumphs. I can recall thousands of people lining the streets for our homecoming and marvel at the millions of dollars raised. What I'll never forget is that the journey unfolded because of the amazing energy of thousands of individuals like Dave who played an important part in manifesting a positive outcome for us. Similarly, when we enjoy a dose of honey, be it drizzled in our tea or smoothed over a peanut butter sandwich, it's easy to lose sight of where it came from.

Life is an intricate tapestry of events that connects us to the greater world around us. Dave approaches the challenges of becoming an accidental apiarist with humour and thoughtful observations. His story reveals the complex and vital roles that bees play in maintaining the delicate balance of a vibrant and healthy planet. Bees, while small, are mighty clever in the way they collaborate to make a difference to our environment. If we look closely, the world is made up of many difference-makers like Dave and his honeybees. Dave's story of his bees is a reminder that we don't achieve anything on our own. Our success is dependent on actions, big and small, of many. Like bees, we all have an important role to play in making this life worth living.

—Rick Hansen

PREFACE

Beekeeping is a sticky, time-consuming, and expensive hobby that often yields little to no honey after carrying out many laborious beekeeper tasks and getting stung so many times you end up feeling like a worn-out, frayed human pincushion. In practice, I have turned out to be a lousy beekeeper. As with so many things in life, I kind of accidentally stumbled into it. I suppose you could say it chose me.

My sister Miriam and brother-in-law Len have been serious beekeepers for years. As is typical with my sister, she took up the hobby with intense focus and vigour. At family dinners, the conversation often turned to drones, queens, nurses, guards, and female workers. Sometimes we would talk about bees too. Then I met my girlfriend, Jeannie, and coincidentally her sister kept bees. Shortly after we began dating, Jeannie started keeping bees herself, and before I knew it, everyone in my immediate circle was decked out in white costumes with wire mesh veils. A cliquey club of honeybee nerds surrounded me, and, all of a sudden, I was the clueless outsider asking dumb questions. I felt alone and excluded—like a bee unable to get into the hive.

All of that changed one spring day a few years ago when I became an accidental apiarist. It happened by chance, and, from the start, I have displayed a tendency to be clumsy, absent-minded, and sloppy around my bees. It's a miracle my girls are alive today. Well, sort of alive.

I've made dozens of dumb beekeeping errors and oversights, which this book will explain in ignominious detail. But along the way I also carefully observed my hive, took notes, and tried to learn from my mistakes. I soldiered on. And I learned to appreciate that being an apiarist offers incredible insight into what bees—an important linchpin in our planet's delicate ecosystem—can accomplish daily through their highly advanced and well-ordered instinctive collaboration. I cannot say I found the same joy in my official studies of the apiarist craft. I took classes and found the material terribly dry and boring, and the same goes for books I picked up at beekeeping conventions. The truth is, I mostly just looked at the pictures and never made it too far past chapter 2. I have been a part-time college instructor for 30 years, and I find most concepts are best learned when enticingly wrapped up in humorous short stories. Mixing some drama and quirky details in with the important, necessary facts increases engagement and helps us remember key points. But maybe I should have finished those "boring" books after all. While I was writing this book, Jeannie harvested over 1,000 pounds of honey and Miriam and Len harvested 400 pounds. What was my yield? I will let you read on to find out.

I wish I could say it doesn't matter to me, that I'm only in it for the good of the earth, but I am more aligned with the famous expression from the 1996 sports movie *Jerry Maguire*, where the NFL football star, played by actor Cuba Gooding Jr., loudly proclaims to Tom Cruise, "Show me the money!" Except I will exchange the *m* for an *h*. If I am going to invest significant dollars and hundreds of hours of my time in a hobby, I want a return. I want something sweet to smear on my blueberry muffin as a reward. Call me a capitalist apiarist or an entrepreneurial entomologist, but "show me the honey!"

Jeannie claims I don't love my bees enough. She says I am not in tune with their needs, that I'm too focused on the honey. I disagree and say, "Don't sweat the small stuff." Hey, wait a minute! All of my bees are small. And worrying about them has caused me a few pounds of sweat.

We do agree on one thing, however: I enjoy writing about bees almost more than I enjoy keeping bees.

WAGGLE DANCING ON A TIDAL ESTUARY

I am not a conventional honeybee expert. Let's get that straight from the start. There are so many fascinating facts to learn about bees and beekeeping that it would take a lifetime to become an expert. For starters, how do these fuzzy yellow-and-brown bugs know where to find the floral cornucopia that yields the sticky pollen they collect on their spindly, hairy legs and the nectar they suck up through their proboscises (a fancy word for bees' tiny nectar-sucking noses)? More importantly, while loaded down with this bounty, how do they find their way back home to their well-organized, factory-like beehive to process the nectar into honey? I have discovered that the hive functions like a harmonious commune with female worker bees collaboratively carrying out an array of specific tasks, and that bees have a remarkable ability to not only adapt but actually thrive after a serious curveball is thrown their way. I will confess that through my beekeeping ignorance and short attention span, I threw my girls many curveballs. Consequently I've discovered those jars of honey we reap from bees are nothing short of golden miracles.

Before I tell you about my initiation into the secret world of bees, my unintentional adoption of 15,000 sensitive, needy insects, and that fateful day I reluctantly first donned the baggy white astronaut suit with the screened hood, you need to know a few important details. One is that I live alone on an old, sea-worn wooden barge. Seventy-five years ago, my barge was used to drag sawdust up and down the Fraser River in British Columbia. Eventually, someone converted the boat into a four-bedroom floating bachelor palace. My 25-by-50-foot red houseboat leaks and it's drafty. Sometimes, depending on weather or tide, it lists to one side. To add to the fun, my water pipes often freeze in the winter, leaving me without water for days. But this houseboat affords me one of the most amazing and breathtaking views on Planet Earth; it floats in front of a small, uninhabited group of islands on a quiet riverbed, a pristine spot that is off most people's radar. I am alone here and no one bothers me. This bucolic setting on the Fraser River is conveniently located only 20 miles from the city of Vancouver, where I used to work as a sports marketing executive with the NBA and the Olympic Games. One important detail is that the elevation of my home rises and falls with the tides, often grounding in the grey, silty mud when the tide goes out. One more thing: the barge has just enough room on the lower back deck to fit a hive of bees.

Whenever my sister Miriam came out to visit me on the houseboat, she would marvel over the lush vegetation on the riverbank and the wide variety of bushes, flowers, and plants on the islands 300 yards from my dock, as well as the miles and miles of crops lining the narrow two-lane farming road leading to my barge. Raspberries, strawberries, and blackberries are cultivated in huge fields across from where I live. Miriam thought it would be honeybee heaven out here, a veritable all-you-can-eat smorgasbord of pollen and nectar.

"Dave," she asked me one day, "how would you feel if we brought a hive of bees out here? I think they would thrive living on your back deck. Len and I will take care of them, so there won't be any work for you to do."

Always game to try new things, I loved the idea. Especially the no-work-for-me part. Having a hive at my place would allow Miriam, her husband, Len, and me to spend more time together. And since I was about to semi-retire from the business world of pro sports, having a new interest suited me. I pictured bees happily flying throughout my river neighbourhood, spreading joy and pollen, while I slathered fresh honey all over my whole-wheat toast and watched the rising sun from the back deck of my floating palace.

As Miriam was selling me on the idea, she gave me my first beekeeping lesson. She explained that bees fly up to three miles from their colony to gather nectar where the gathering is good. Bear in mind that these little creatures' brains are one twenty-thousandth the size of ours. But despite brain-size limitation, and without the aid of a GPS, bees have the uncanny ability to return to the exact same spot after every single flight, to the hive they originated from, which in this case would be on the deck of my houseboat. Even with my relatively giant brain, I often can't remember where I parked my car at the local shopping mall.

Bees, my sister went on to tell me, can also travel up to a remarkable 15 miles per hour on the equivalent of a mini two-stroke engine. The two strokes that propel them are a fragile pair of tiny wings that flap 11,500 times per minute.

Miriam stressed that honeybees care about only two locations when they are out flying around: the nectar and the hive. What I liked most about my first primer on bees was the whimsical name for how bees communicate to one another the direction and distance to these two important locations. This über-scientific beekeeping term is the *waggle dance*. Bees leave the hive every day to seek out abundant nectar flows; when they return to the hive, they need to let hundreds of other bees know where to go. Since they don't know how to talk, they dance to communicate.

The waggle dance is a highly specialized and sophisticated bee boogie that points the way to robust nectar flowing in plants and flowers in the neighbourhood. The complexity and effectiveness of what happens when waggle-dancing bees get down, get funky, and bust a move in the hive are truly unbelievable.

Here's how it works. While the worker bees are out gathering nectar, they rely on solar navigation to pinpoint the best blooms. Each bee cross-references the exact position of those plump, juicy flowers with the sun's location in the sky at the precise time she is there. She memorizes the sweetest spots within three miles of the hive and then produces a mental road map that will direct her sisters to the floral treasure. After she returns to the hive, the real work begins on the dance floor.

Once inside the hive, she spins in circles and waggle dances in an elongated figure-eight pattern with the axis tip pointing in the direction of the blooming flowers. This visual is tough to describe; type "waggle-dancing bees" into YouTube to watch it. As she dances, hundreds of her fellow hive sisters gather around to touch her, crowd into her, and rub against her, absorbing and interpreting her frenetic energy. They are memorizing her report of the exact direction of the best flowers and details about how to get there. The length of time that she waggles communicates the distance to the flowers. The longer she waggles, the farther away the flowers are; the shorter her waggle dance, the closer the flowers are. One second of waggle equals about 0.6 miles. How the heck scientists figured this out is beyond me. I had to see it to believe it. So one day my sister invited me out to her place, where she pulled a frame of bees from one of her hives. I stood back amazed as together we witnessed these disco-dancing bees getting down. In her white protective suit, Miriam held out a single frame of cells with hundreds of bees crowding and pushing to get in close to the one extremely popular and knowledgeable waggle-dancing bee. Miriam told me that once the bees interpret and memorize the waggle dancer's instructions, they go and find those exact patches of blooms, as well as other good blooms along the way, and, in turn, pay it forward. Some of them waggle dance when they come back, and the beat goes on; they waggle on, and the information sharing minimizes the energy the hive collectively invests in foraging missions. Bees are collaborative, and nowhere is that trait more evident than in the waggle-dance process.

For bees, the waggle dance has very little to do with dancing and everything to do with survival. But, of course, it reminded me of learning some of my best dance moves watching *American Bandstand* and *Soul Train* on television in the early '70s. Remember how Dick Clark would interview couples after each song and ask them to rate the tune? After dancing to the latest hit, I never heard the answer "I really liked the beat. The song was easy to dance to, and by observing how my partner moved in circles and flapped her arms, I can tell that there are some really prime begonia blooms 1.5 miles south of here at the intersection of Hollywood and Vine." The teenagers dancing on those shows may have been looking to get some honey, but it was a different kind.

At the same time that Miriam was introducing me to the waggle dance in preparation for her hive's debut on my back deck, my girlfriend, Jeannie, was busy with her own new beehive and consulting with her own beekeeping sister.

Jeannie kept her bees on the lawn of her lovely half-acre property, which she occasionally rents out as a vacation home. After acquiring her hive, she discovered that some guests are not fond of bees. I suggested that when she has bee-phobic guests, she simply move the hive to a more secluded location on her property. She then taught me one of the countless guidelines that seasoned beekeepers know: Always keep your hive in the exact same place. Don't ever move the bees. The general rule of thumb among apiarists is that beehives can be moved only about two feet. Oddly, they can also be moved a distance of two miles or greater. However, moving them any distance in between those two measurements messes with their internal GPS units. Here is the rub: the bees get used to their hive being in a certain location, and all of their finely tuned waggle-dance flight paths are programmed from that location. Thus, if you move the hive a few inches, it is no big deal. They will find it. But if you move the hive 15 feet, let's say, they will get lost as soon as they leave the door to forage for neighbourhood pollen. And since they don't scatter bread crumbs behind them like Hansel and Gretel to find their way back home, their return voyage will be hampered as well. Studies have shown that if you move the hive a few miles, the bees quickly reprogram their adaptable bee brains to accommodate the switch. I told you they were smart.

Jeannie once moved her hive 20 feet, and when her bees returned home, they went to the exact same spot the hive had been before they took off. That afternoon, she noticed a large swarm of confused bees flying around in the air 20 feet to the left of the hive. It prompted her to move the hive back.

The significance of the waggle dance and the importance of not moving the hive were the first two nuggets of entomology knowledge—and certainly not the last—I didn't pay enough attention to. I would go on to forget, misinterpret, and disregard hundreds of other essential beekeeping facts as I struggled to become an apiarist. Overlooking these first two pieces of knowledge, though, would cause me a great deal of angst on the seminal night my sister and brother-in-law brought the bees out to the barge.

We are all likely familiar with the moon's effect on large bodies of water. In my case, that body of water is the mighty Fraser River, with its headwaters up near Mount Robson, the highest point in the Canadian Rockies. The Fraser River begins as a fast stream of clear, pristine mountain water. It winds its way through British Columbia and picks up more and more water as creeks, streams, and other tributaries empty into it. By the time the Fraser reaches Ladner, British Columbia, where I live, the wide, silty river is more accurately described as a tidal estuary. It is so close to the Pacific Ocean that the daily tides dramatically change the river's height—up to 15 feet each day.

Many unsuspecting dinner guests have ventured out to my old barge, often arriving around 5:00 PM. They stroll casually down a gently sloped ramp from the riverbank to the dock. You can imagine the surprised looks on their faces when they leave at 11:00 PM and have to ascend the same ramp, now in a steep Mount Everest climb position because the river dropped 15 feet while we dined. It can be a tough scramble back up to the land, especially after a few glasses of wine. At low tide, brown mud replaces all of the sparkling water that surrounded my boat. Most people appreciate that my home sways a bit with the current, or when a big fishing boat goes by, or when the wind blows, but they are unaware of the strong tidal forces that turn my home into an elevator constantly moving between the first and third floors. Recall the apiarist rule of thumb: the hive can be moved only two feet or a few miles.

Up until the evening my bees were to arrive, I hadn't put two and two together. What effect would the rising tides have on my waggle-dancing bees? I didn't make the connection. It's not as though I didn't think a lot about the hive on the houseboat. But I thought of questions like: Where on the back deck would we put it? Would the bugs survive the strong westerly winds on the river? How would my neighbours feel about the arrival of tens of thousands of insects with sharp barbs on their butts? I wondered if anyone in my neighbourhood was allergic to bees. I worried that the bugs would create a disturbance flying around outside, annoying my guests when I hosted dinner parties. I never thought of the fact that my back deck would become a moving target with a 15-foot range for my tired, homeward-bound bees just looking for a place to drop off their pollen and nectar load and lapse into a dreamless sleep. This tiny oversight was rather significant; the bees were currently in a van travelling down the highway heading for my houseboat on the tidal river.

The movement of bees in a van across town is an art in itself. It goes without saying the hive needs to be carefully sealed—really carefully sealed with strong grey duct tape. If you think driving while on your cellphone is a distraction, try driving with hundreds of recently escaped, angry, and confused bees obstructing your vision. When you are transferring bees to a new location, it is best done at night, when the bees are more passive and all of them are at home. If you move the hive during the day, you could leave behind a large percentage of bees that are out foraging. Relocating hives is a nocturnal activity.

When the evening of the big move finally arrived and Miriam called me from the road to say they had the new beehive tightly sealed and secured in the van, it still hadn't occurred to me how detrimental the tide's movements might be to my new six-legged houseboat residents. Why it hadn't occurred to my sister, I still don't know. Maybe she thought of it and forgot to tell me; or maybe she knew about the tide differential and thought the hive she was bringing was super smart and advanced and up for the challenge. After I hung up the phone, I relaxed for a few more minutes on the couch with Jeannie and had another sip of wine. Then, suddenly, it hit me.

"Oh no," I exclaimed to Jeannie, who sat with her eyes closed and her head tipped back on the couch, looking rather peaceful. "I just realized . . . I don't think this is going to work! The tidal movement out here is really going to mess with the bees' navigation instincts." Jeannie sat up abruptly. She knew exactly what I meant. We stared at each other in silence. I immediately called Miriam back and told her of my potentially fatal oversight. "Miriam, stop!" I shrieked. As a novice apiarist, I got a little overexcited. "Don't bring the bees out here! The way my houseboat rises and drops 15 feet each day, there is no way they can survive!"

There was a very long 10-second pause. That rarely happens with my sister. When she came back on the line, we discussed the predicament at length, eventually deciding that since she was well on her way, and since she and Len were already wearing their beekeeping gear and had sealed the hive, she might as well just come on out and we would stick with the original plan.

As we nervously waited for the 15,000 bees and my sister and brother-in-law, Jeannie and I discussed how beekeepers try to replicate a natural hive environment. Most wild hives are located in trees or old logs on the ground. The stationary wooden homes we build for bees, called Langstroth hives—many of which look like an old-fashioned, four-drawer filing cabinet—simply try to replicate a hollow space bees might find in a natural setting. But nowhere in nature would bees choose to build a hive on a river. Or, even worse, a tidal estuary. Would a log floating down a river ever be a place where bees would start a hive? No way. This could well be one of the first times ever that honeybees would be raised on top of a river. To clarify my concern, I came up with an aeronautical analogy.

"Jeannie," I said, "let's pretend you are an air traffic controller giving directions to 747 pilots on their individual flight paths. The planes you are directing fly all over the world and need precise instructions on how to get to their exact destinations and back. Unlike a bee, as an air traffic controller you get to actually speak to the pilots and don't have to waggle around in figure eights, flapping your arms. Unbeknownst to you, the runway you are dispatching the planes from is moving up and down the whole time. You'd have planes crash-landing left, right, and centre."

Jeannie and I arrived at the same conclusion as we pondered this crazy predicament: the bees in the van were heading for big trouble. If we thought we were screwing around with their heads by moving them across town, just wait until they arrived at the boat. Were we condemning them to a fruitless, empty life of constantly searching for their shifting hive? I hadn't yet even taken delivery of the bees and I couldn't get three detrimental adjectives out of my head: homeless, helpless, and starving. Would the local apiarist community come to know my houseboat as a slave-worker death camp full of directionless bees? But no matter how much Jeannie and I worried that we had made a big mistake, there wasn't much we could do about it now. There was no stopping what could be one of the biggest biological blunders in beekeeping history; Miriam and Len were just down the road, and we had all agreed to go ahead with the half-baked original plan. We just had to hope the bees would figure it all out on their own once they got settled in.

Miriam and Len arrived shortly after 10:00 PM looking like a couple of aliens in their white bee suits. Transferring the bees from their van parked at the side of the river to my back deck took half an hour. We very carefully carried the beehive boxes through my kitchen and living room. During the indoor portion of the transfer, I saw one bee escape from under a fraying piece of tape and land on the counter by my sink.

After the hive was securely in place outside on the deck, Miriam and Len removed the tape from the hive's small entrance, leaving the bees free to go out and forage the next morning. I thought to myself, "Girls, after you leave here tomorrow on your nectar extravaganza, good luck finding your way back." Then a more macabre thought came to mind: "Rest in peace."

With the transfer taking place late at night, and Miriam and Len in outfits concealing their true identities, the whole evening felt like clandestine skulduggery, especially with the knowledge that genocide was a possible outcome. I was glad it was dark and no one knew what we were doing. If this ill-conceived beekeeping exercise flopped, I worried I would have an animal welfare inspector knocking on my door with a search warrant.

After we put the bees to bed and left their front door wide open, we were clearly well beyond the point of no return. The hive, now firmly placed in its new marine location, had 15,000 bees cozily sleeping the night away. Before going back inside, I observed that a few of the bees ventured out of the hive for a midnight snack; I supposed they couldn't sleep due to all the jostling and being in an unfamiliar place. I was sure they were a little disoriented and hoped they'd return. With all the heavy lifting done and the excitement subsiding, I invited my sister and brother-in-law to stay for a cup of tea with—you guessed it—a big dollop of honey. As the four of us talked inside the float home, it became clear we were in a bit over our heads and needed to seek advice from beekeeping experts.

After Miriam and Len had changed back into their street clothes and left, I was doing the dishes when the single escaped bee flew up and stung me on the arm. My first sting. Was it some kind of ominous sign?

In the days after they dropped off the bees, Miriam and Len canvassed the bee intelligentsia for answers to our perplexing question: Could bees survive the daily 15-foot elevation gains and drops on the Fraser? Miriam's bee club was full of knowledgeable beekeepers, but this question had never arisen before. The experts were stymied. The province of British Columbia has several regional bee inspectors whose job it is to help beekeepers raise bees. When she asked these various officials about our predicament, there was varying conjecture, but no one really knew for certain.

I carefully observed the bees for the next few weeks as I walked up and down the ramp to my dock. There were always a couple dozen bees hovering around the hive's entrance. I took this as a healthy sign. On my front upper deck, where I often recline in the sun, I only occasionally noticed one or two bees buzzing about, and they never became annoying. The bees seemed to dwell on my houseboat peacefully, never once bothering me. I guess the foraging in the fields and on the islands was just too good. I did have the occasional errant winged visitor inside. In the past I generally killed any bug that had the audacity to come inside my float home. But now, if I identified the bug flying around my kitchen as one of my girls, she got an immediate stay of execution. I was beginning to like bees.

About six weeks after dropping off the bees, Miriam returned to my place in her white costume and did a check on the hive. She was amazed at how well the bees were doing. The hive showed no signs of the parasites that often invade and spell the beginning of the end. That day in early June my hive got a clean bill of health—no mite parasites and no slimy diseases, just beautiful new cell formations, plenty of worker bees, and a busy, happy, plump queen planting baby bees into the cells. Best of all, the bees were making lots and lots of sweet, delicious honey. The universe was unfolding as it should in this hive. The girls were not getting lost in their new surroundings; they were finding their way back home.

Miriam came out again in July. This time she couldn't contain her excitement as she opened the lid of the hive. "Dave!" she shrieked. "I can't believe how much honey you have in here and how perfect the hive is!" She told me my hive might produce as much as 100 pounds of the golden elixir—more honey than any of her stationary land-based hives had ever yielded. How exhilarating! Just a couple of months earlier, I had thought our irresponsible hive placement had condemned the girls to a struggling homeless existence at best, and the Grim Reaper at worst.

Miriam and Len came out a third time in September, this time to remove the hive's honey frames and take them home to extract the honey. I should have called a Brinks armoured car, because we hit the honey jackpot: 100 golden pounds of amazing syrup stashed away in the hive's tiny cells.

Miriam and Len extracted all that honey in September. In October they filtered the honey and bottled it, taking us to what can best be described as the Academy Awards for British Columbia honey producers. Every other year, hundreds of members of the British Columbia Honey Producers' Association have an annual meeting. One very important item on the agenda is to choose the best honey in the province. Taste is an important attribute, but it is just one of the many categories that a judge weighs. Each jar of honey entered in the prestigious competition is examined against stringent criteria, including air bubbles, density, moisture content, clarity, brightness, and lack of dust particles. Even fingerprints on the jar are a factor. And . . . drum roll, please . . . cue the trumpets . . . the province's leading honey experts ranked our Houseboat Honey as the second-best amber honey in all of British Columbia!

When Miriam called me with the news, I raced out to the back deck to congratulate the girls and spend some quality time with them. I was so proud of each and every one of them. As I sat beside the hive, I reflected on what they had accomplished. They had not only adjusted but *thrived* in their new hive location. They'd adapted to new surroundings, new foraging grounds, and a constantly swaying, moving, rising, and falling home—the first of many curveballs I would throw them. I told them, "Girls, although I am very happy with the way you overcame the adversity of the rising tides and went on to produce 100 pounds of amazing honey, we did only come in second place. Now get some rest, try to keep warm over the winter, and next year we are going for *number one*! Girls, our motto out here on the float home is 'Sisters Are Doin' It for Themselves'!" As I went back inside my houseboat to put on the kettle, I thought, "This beekeeping isn't so bad after all."

That Christmas I got a card from Miriam proclaiming in her beautiful calligraphy that the hive was officially mine. The drive out from her home to the river to tend the bees had proven to be too long. And besides, I had shown some interest in the bees, and it was time for me to take that big step toward tending my own hive. "Don't worry," she said. "We will always be just a phone call away, and besides, Jeannie can help you too." So much for the no-work-for-me part. But I was excited to learn the ropes. I knew the hive would soon grow threefold over the summer to 50,000 bees. That's a lot of little creatures for which I was now 100 percent responsible. Like the father of a newborn baby, I was up for the challenge. There was so much to learn, so much to prepare for. A new dimension of purpose and meaning had been introduced to my life, and things were looking good. If the Houseboat Honey haul was any indication, I was about to become a successful beekeeper with honey to spare.

Anyway, I had no choice. The proclamation on the Christmas card was crystal clear. The transfer of ownership of the hive was complete: signed, sealed, and delivered. The bugs were mine. I had to sink or swim. So for the next few years I did everything I could to learn how to keep a beehive alive.

THE STING

Bees sting. Beekeepers get stung. That's just the way things are.

The road to becoming a true beekeeper, I was quickly learning, is itchy and sometimes painful. You must respect your bees and be at one with your hive to avoid pain. Occasional stings are inevitable. But how will you know when you have transitioned from a maladroit beginner like me, clumsily poking through your hive, to a skilled and sensitive apiarist assisting in the hive's overall well-being and providing optimal circumstances for honey production? By that I mean embracing all facets of beekeeping in your heart and soul and respecting the bees for what they are: incredible, industrious little beings who have a right to be here, who truly coexist with us. Though I've not yet reached the level of "skilled" apiarist, I'm getting closer. I've gone from swatting at bees to embracing them (well, not literally). I am a 195-pound man who now feels genuine compassion and sympathy toward an insect that weighs only about one-tenth of a gram. One could say the tiny beasties have taught me a great deal, changing me in remarkable ways.

I sometimes wonder how different beekeeping would be if the little creatures didn't sting. First of all, I am sure more people would keep bees. Also, an entire chapter of *Winnie-the-Pooh* would have to be rewritten. And finally, beekeeping would not be as much fun because you wouldn't have to wear the white head-to-toe suit. Being an apiarist allows you to dress up in a crazy costume, complete with headgear and leather gloves, and celebrate Halloween year-round. But we don't wear the costume for fun; we wear it, of course, to protect us from the bees' barbed stingers. Stings and the ensuing pain, caused by a bee's melittin venom, are as much a part of beekeeping as delicious honey. It's the perfect yin-yang relationship. Hurt and suffering must be endured before you can truly appreciate and enjoy the sweet, golden reward.

If bees didn't sting, then bears, wasps, skunks, and humans would inevitably steal all their honey. Bees produce a valuable commodity that many of the creatures roaming this planet, including me, want. Somewhere in the constitution of nature, bees were granted the inalienable right to bear arms.

The fact that bees do sting forces beekeepers to be more observant, more *in the moment*. I have a distracted focus at best; if I were a schoolboy in these modern times, my report card would no doubt say in bold letters, "Dave has trouble paying attention." Nowhere is a lack of focus more immediately punished than in the hobby of beekeeping. Each time we go into the hive, 50,000 loaded stingers await us.

The minute you open a hive—no, even as you *approach* the hive—you need to be completely aware of what is going on around you. No daydreaming or wandering off in thought. Hurried actions, abrupt movements, or anxious energy can turn a peaceful commune of honey makers into an attacking squadron of angry, venom-spewing kamikaze pilots.

And just like how the kamikaze pilot missions in the Second World War were suicidal, a bee is in suicidal mode when it stings you. It will die within an hour because the stinger, which lodges in the victim's skin, tears loose from the bee's abdomen, taking its guts with it.

One time I carefully caught a bee who was trespassing in my float home in my leather glove. Before I let it go outside, I examined it under a magnifying glass. I gently held her between my gloved fingers, careful not to squish her delicate frame. To my horror, as I drew the magnifier closer, she buried her little rump into the glove's thick leather hide. "No, no, don't do it!" I cried. It was too late. I slowly spread my fingers as she crawled up my thumb dragging a gooey elastic trail of guts. To be more exact, her digestive tract plus muscles and nerves trailed in a gross three-inch string. Yuck. There was nothing I could do; I knew she was headed to honeybee heaven. So I stuck her under a glass on the kitchen counter, gave her some sugar water, dimmed the lights low to comfort her, and played Beethoven's Symphony No. 9 in Bee Minor for her. Within an hour, sadly, she died.

Many bees have sacrificed their lives just to plant a stinger deep into my skin. Bees place protecting the hive ahead of life itself. Admirable? You bet. Especially when you consider how dangerous it is for bees to approach a human. Their loud buzzing makes them easily swattable targets. Before becoming a beekeeper, I batted many an angry bee onto the ground outside, where I firmly planted my foot on its prone little form to crush it. Wasps are even worse.

Indoors, I had more elaborate ways of killing bugs, including bees. An old-fashioned fly swatter, a novelty battery-powered electrified bright yellow plastic tennis racket, or a simple rolled-up newspaper were my favoured methods of murder. I played judge, jury, and executioner. The bugs were always found guilty. It's hard to fault anyone for swatting an attacking bee. After all, their stings hurt like hell, and bee stings release pheromones that prompt nearby bees to join in. Get stung by one bee and 100 reinforcements could well be on their way.

A bee sting may also trigger anaphylactic shock, which can be extremely serious. My girlfriend, Jeannie—a competent, cautious, and caring beekeeper—once had a terrible allergic reaction. Like most beekeepers, she had been stung several times over the years. Those early stings resulted in nothing more than agonizing pain. But after three stings in one day, she had to be rushed to the hospital, where the doctor ordered her to stay overnight.

It was good the sting occurred while she was tending her own bees, because she was with her sister Suzanne, who is also an experienced beekeeper. Immediately after the piercing venom took hold, Jeannie knew something was seriously wrong. She and Suzanne wasted no time getting in the car to hurry to the hospital. As they embarked on the 10-minute drive, Jeannie's heart began beating faster, her skin became uncontrollably itchy, and her whole body heated up. Three or four minutes into the trip the symptoms became unbearable. Jeannie spotted a pharmacy, told her sister to stop, and dashed inside. Grabbing a package of Benadryl, she ran to the counter to pay, then downed a tablet. Benadryl blocks the effects of histamine, a substance released by the body if you are allergic to bee stings.

Back inside the car, Jeannie was in sheer agony as her condition escalated. Pink welts the size of hockey pucks broke out all over her skin. She reported that her heart was beating twice as fast as it should, pounding like never before, and she overheated to the point where Suzanne could feel intense heat radiating off of her sister. As they sped down the street, Jeannie had to rip her shirt off her raw, red, and raised skin. In her crazed state, she hung her naked torso out the passenger window of the speeding car, drawing temporary comfort from the cool breeze.

Luckily, Suzanne had an extra-large, loose shirt on hand, and Jeannie put it on for the last few minutes of the trip to the emergency ward. When they got there, Jeannie sprinted to the smooth Arborite check-in counter, where she unabashedly lifted up the shirt to place her bare chest and stomach against its cool surface. The emergency doctor arrived to witness a writhing, bright red, overheated woman in intense pain and declared he had never before seen a bee sting inflict such a severe skin reaction.

After a long, sleepless night, Jeannie was released the next morning. Thank goodness she was back to normal within a few days. When we talk about it now, she repeatedly points out how it all happened so quickly, and I can't get over how such a minuscule amount of venom from a bee's stinger can create such massive biological havoc. To this day, she visits an allergist monthly for preventive treatment.

One of my girls once nailed me in the face. It hurt, but after 15 minutes the soreness went away. I had forgotten the entire incident by the time I went to bed that night. The next day I was due to go to a function in Vancouver with officials from the Beijing Olympic committee. I was aghast when I woke up and couldn't see properly out of my left eye. I got out of bed and looked in the mirror—my eye was swollen to a narrow slit.

Is there any antidote for bee stings? Well, sort of. Pharmacies sell pen-like dispensers containing a balm meant to be applied to the skin immediately after a bite. I have found a moderate degree of relief from them. My sister has good things to say about a poultice of baking soda and water. But my favourite nullifier comes from a distant childhood memory and a Greek family that lived down the lane from us when I was about five or six. Mrs. Michelidies told me that if a bee stung me, I should immediately pee on the soil near the hive, then bend over and pick up the pee-soaked mud and apply it to the sting. I talked with my sister Miriam about this one, comparing it to her baking-soda poultice. As I was laughing, Miriam gave me pause when she told me there is science to back up this remedy: urine can effectively treat all stings. When Miriam and Len were at a resort in the tropics, a jellyfish stung her while she was happily snorkelling in the green-blue water. As they returned to shore, fellow snorkellers offered to pee on her arm! Was this an example of well-meaning tourists wanting to help, or some kinky sun-baked weirdos looking for perverse kicks?

At this point, you may be wondering if my lack of focus impedes my ability to remember to put on a beekeeping suit. Believe it or not, although the bee suit offers some degree of protection, bees can invade the tough, canvas-like material. It is incredible that a bee's tiny stinger can permeate such thick fabric; maybe they use a pencil sharpener on their stingers, or maybe they carry mini seam rippers. And if you don't seal your bee suit properly when you put it on, errant bees gain access through the bottom of the leggings or the open ends of the sleeves. Try having an angry bee fly up through your leg into your crotch. It's almost enough to make me switch hobbies and take up stamp collecting.

So beekeepers must approach their hives with proper preparation, in a relaxed manner, and with a degree of caution. As I put the bee suit on, I often inhale deeply several times to try to slow my breathing down. I envision the hive as 50,000 happy friends under one roof. If that isn't a bizarre enough fantasy, I go on to visualize how welcoming they will be when I lift the lid off of their small home and pay them a visit. The breathing exercises and the role-playing take the edge off what can be a nerve-racking experience. I tend to be less clumsy when I am more relaxed and focused. Without my preparatory ritual, I am inclined to drop tools, knock things over, and generally annoy and stir up the bees. Jeannie, who has far more experience than I do in the hive, is always methodically slow and confident in her approach.

It's important to keep in mind that no two hives are the same. Each hive has its own culture and "personality."

Some hives consist of passive, good-natured bees that cooperate in commune-like bliss, while other hives are more like a collection of jail cells with angry inmates ready to violently attack intruders and escape. The time of day should also be taken into consideration when approaching the hive. It is best to open up the hive to make a house call when many of the bees are out foraging for nectar—early afternoon on a warm sunny day. Half the bees won't even be home, and the ones that are will be so busy storing the harvest and building honeycomb that you will hardly be noticed. Theoretically.

Considering the potential peril of approaching and entering a beehive, you may be asking, "What is it that beekeepers are actually doing in there?" Well, about half a dozen tasks require "looking under the hood." We go in to check the development of comb, ensure the queen is alive and producing eggs, and watch for minuscule enemy invaders, such as mites. We lift the inner frames to determine how heavy each one is as an indicator of how much honey has been produced. Sometimes, when the flowers are not making enough nectar, we go in to feed the bees sugar water.

When I feed my bees, the unique location of the hive on my float home's back deck allows me to stay safely inside my house and simply push the sliding window over four inches. This gives my hand just enough room to slip a jar of sugar water into the hive. To do this reasonably cautiously, I wear the leather long-sleeved gloves that came with the bee suit and put on the hood. This method works well most of the time, but invariably the occasional bee will fly inside. I found myself relaxing the hard and fast rule I'd had prior to becoming a beekeeper: any bugs inside my place were trespassing and, as such, I had the right to kill them. Bees ceased to be annoying intruders and transformed into something between a carefully tended crop and a herd of beloved pets.

In life, what you do when no one is watching is what matters the most. One day during a feeding, I happened to leave the window open a bit too wide, and about eight or nine bees flew into my home. I was in harm's way because I hadn't bothered to put on my bee suit. It took me 15 minutes to track down each and every girl and gently escort her out the front door. The best way to achieve this was to gingerly caress each wayward bee in my glove, ever so careful not to squish her. I did this task with such a degree of caring that it surprised me. Six months prior, I would have just rolled up a back issue of the *Vancouver Sun* and chalked it up to self-defence.

As I released the eighth unharmed bee from my front deck and watched her fly around back to the hive, I thought, "There is hope for me." The transformation had begun: I was slowly turning into an apiarist.

THE PROCESS OF CREATING AN ALMOST-PERFECT FOOD

Created from fields of wildflowers by remarkable flying insects in summertime, sold in pretty jars at farmers' markets, and consumed blissfully in steaming cups of herbal tea, honey is regarded as the quintessential natural food. While raw honey is technically unprocessed, it does go through quite the process to get to your pantry. And the result is a product that's almost flawless. In fact, it's about three-quarters perfect.

In general, food should taste good, look nice, be easy to serve, and have a decent shelf life. Humans have turned processing foods into a massive industry to try to nail this elusive equation—as evidenced by the popularity of overly processed foods like doughnuts, corn chips, pop, and hot dogs. Honey, however, comes by three of these attributes honestly. Admittedly, it does leave a lot to be desired in one category.

Honey's sweet, sunny flavour simply cannot be improved upon. Period. Don't mess with Mother Nature—no additives or flavour enhancement required. Honey is one of the most delectable foods on earth, dependably sending the brain's taste receptors into an orgasmic frenzy. I routinely apply honey much too thickly on my toast, I drop spoonfuls into my coffee, and sometimes I just smear it on my fingers and lick it off.

Translucent and smooth, honey is beautiful to behold. The light orange-yellow liquid glows; it is lazy and slow, like a summer afternoon spent dozing in a lawn chair. Sometimes I hold a jar of it in my hands and lift it up to a window just to bask in its golden magnificence. I have passed the time by turning a newly sealed jar of honey upside down to watch air bubbles slowly and gracefully rise back up to the top of the jar, like a lava lamp.

When it comes to honey being easy to serve, however, all I can say is that it's hard to be perfect. Honey scores a zero in this category. I don't know about you, but no matter how careful I am, it always leaves a sticky mess on the counter, around the jar lid, on coffee cups, and sometimes on my fingers and chest. I've tried dispensing it with those old-fashioned wooden dollop sticks (honey dippers); they don't work. Then there are the bear-shaped plastic squeeze bottles. They had great potential as stickiness-free honey dispensers, and I joyously refilled them with my own Houseboat Honey. Then one morning I reached into my cupboard and wrapped my hand around the bear's soft plastic belly only to have my fingers slide and stick on a layer of honey that had dripped down its torso.

While stickiness is one drawback, honey also has a bad habit of changing consistency, based on temperature, moisture in the air, and age. How many times have you bought a nice jar of amber liquid and then left it on your counter for a few months only to discover it has crystallized? I invariably encounter this tragedy at the worst time: in the early morning when I am late for work and dying for some honey on my brown toast. In my semi-slumberous state, I grab the plastic honey bear, twist the bright yellow lid off its noggin, plunge a knife down its throat, and am forced to scrape hard honey crystals from the insides of the bear's chest. Then I mangle my toast, trying to spread it.

Honey does, however, get a five-star rating when it comes to longevity. Honey never goes bad. Ever. Not after a month, a year, or even a decade. Some people think honey is no good after it crystallizes. Wrong. Crystallization is honey's natural process; the sweet taste and exemplary qualities are still there even after it hardens. To transform your crystallized honey back to its original liquid state, all you have to do is put it in a glass jar, submerge it in a pot of water, and heat it up. Please note the *glass* jar. Don't do what I did and put the plastic bear bottle into the pot on the hot stove. I melted Pooh's feet off, and he bled his contents into a toxic mixture of boiling water and melted plastic. I frantically pulled the bear's head out of the pot only to drip sticky honey and plastic all over the floor, the counter, and my suit pants.

There are documented cases of containers of honey buried with Egyptian mummies thousands of years ago. This is no surprise because mummies were buried with their prized possessions, personal belongings, and food to assist them in their journey to the afterlife. The archaeologists who discovered these ancient tombs must have been astounded to find an earthen ceramic pot of honey nestled next to King Tut. It probably took a lot of nerve for them to dip their fingers into the honey, lick it off their fingertips, and declare the honey still fresh. But they did. Google it.

Since honey can last for thousands of years in the oven-hot Egyptian desert, it should come as no surprise that it needs no refrigeration. I am bewildered when I snoop in friends' fridges and find jars of honey. If you are keeping your honey in the fridge, take it out and save your fridge space for Mother Nature's more spoilable offerings, such as milk and cheese, foods that are not as perfectly engineered as ever-fresh bee ambrosia. Store your honey in a cupboard at room temperature and pull it out in the year 2774 to experience flavour as fresh as the first day a forager bee collected it and a worker bee fanned it with its wings.

Honey, without any human intervention, meets three out of the four desirable processed-food attributes—like

I said, it is three-quarters perfect! And bees go through one heck of a complicated process to make it so, carefully gathering, chemically modifying, and storing it. After bees extract the nectar from the plants with their proboscises, they swallow it. I myself was surprised to learn they don't bring it back in tiny buckets. In fact, when bees make honey, it's a cross between a puke-a-thon and a barf-fest. A small portion of the nectar they swallow goes into their stomachs to give them the energy and stamina they need to visit up to 1,500 plants and flowers daily. The rest of the nectar goes into a spare stomach for storage. But wait, it gets grosser. When they get back to the hive, each worker bee finds a processor bee and then regurgitates the nectar sucrose from its storage stomach into the processor bee's mouth.

Here is where the production gets a bit technical. The processor bees that stay in the hive keep the sucrose nectar in their stomachs for about half an hour. During those 30 minutes, the bee's stomach enzymes break down the sucrose nectar by converting sucrose into glucose and fructose; we're talking a highly complicated chemical operation. Processor bees are younger than worker bees, so they still have the proper stomach enzymes to break down sucrose. Once bees get older, they lose those enzymes and have to go out and forage (or work, hence the name worker bee). The analogy is striking. For many of us, our youth is spent drinking and barfing; then as we get older we have to get on with it and go out to work.

After half an hour of stomach-based chemical conversion, the processor bee is ready to hurl the contents of her stomach, and it's not like she can reach for a barf bag with one of her six arms. There are thousands of processor bees in the hive, so there wouldn't be enough barf bags to go around. Each bee just pukes the glucose and fructose concoction from her stomach into one of the handy six-sided cells that comprise the hive. It is not quite honey yet. The solution is too watery, so the processor bees spend a day or two fanning the honeycomb cells with their wings to reduce the water content.

If you were feeling sorry for the poor old forager bees out there in the fields slaving away while the young processors have it easy, think again. Here is the most disgusting step in the process. After fanning the honey, the processor bees must secrete wax out of tiny glands located on the undersides of their rear ends in order to cap each cell. You could say the processors are literally working their little behinds off.

Finally, after all of the bees' hard work producing honey to sustain them over the winter, humans come in like Jesse James to rob their precious stores. The human honey-making process—or, more appropriately, honey-*taking* process—is simple. It's also very sticky and hasn't changed much in 100 years. Did I mention it is very sticky?

Once a hive box is full of honey, you must begin by pulling the individual frames out, much like you might pull files out of a filing cabinet. These frames weigh about a quarter pound each when they are first installed and empty in the spring. At the end of the summer, if the weather conditions are perfect and you are just plain lucky, the frames should weigh about 3 pounds each. A box of 10 honey-laden frames weighs about 30 pounds, and a good harvest would consist of three boxes, or 90 pounds.

After removing the frames from the hive boxes, you need to take a brush, similar to the hand-held kind used to sweep dust into a dustpan, and gently sweep the bees crawling all over the frames back into the hive. Since I rarely sweep my floors, I don't have much practice at this, and I've had to learn the hard way that if you sweep too hard, you crush the bees. If you sweep too lightly, the bees are not fazed in the slightest and keep right on busily crawling about their tasks. Bee sweeping is a fine art.

When the frames are free of bees, you must inspect them to ensure the honey cells are capped with wax. Uncapped cells are a signal that the bees were likely in the middle of fanning the honey to reduce its water content and it is not yet ready for harvest. Can you say "soggy toast"?

Once the frames are removed from the hive, they need to be stored in tightly sealed tubs. Tupperware makes ideal large rubber tubs that most beekeepers use. Don't forget to seal the lids properly or, just like when Jesse James robbed banks and the sheriff pursued him, the bees will chase you down to get their honey back. Leave an unsealed tub of frames near a hive for five minutes and hundreds of bees will try to reclaim it. It is best to make a clean getaway. With proper storage, those frames of honey can be left for years. However, most people extract the tasty ambrosia shortly after the heist.

Honey extraction is a three-step process. First, the sealed hexagon-shaped combs need to be unsealed. This requires an uncapping fork, which looks like a hair pick and might have been used to coif the hippest Afros in the '70s. I like the uncapping fork because it's fun to run the sharp metal comb over the thousands of tiny cells and watch them burst open with honey. Don't forget to uncap both sides of the frame. If you want to use Mother Nature's friend Omnipresent Gravity to get the honey out, leave the frames for a week or two in a very hot room until it all drips out by itself. But most of us don't have time to wait that long, so this is where the honey extractor comes in.

A honey extractor is a round contraption, usually made of stainless steel, about the same size as a washing machine. Four or six frames go into special cages in the extractor that are attached to a hand-crank gear assembly. When you crank the gears, the cages holding the frames spin around faster and faster and faster. When they are spinning fast enough, Omnipresent Gravity's cousin Centrifugal Force arrives to lend a hand. The sheer force of the circular whirling pulls the honey out of each frame. It takes about two minutes of hard cranking for all of the honey to splatter onto the clean walls inside the extractor. The honey then slowly flows to the bottom of the extractor, where a valve opens to let it flow into big plastic pails.

The filtering is next. Most people don't like bee legs, antennae, little pieces of wax, or tiny twigs in their honey. Too many unwanted bits and pieces of debris can turn honey into stew, so the honey needs to be strained through a small wire mesh screen placed on top of an empty tub. Do this in a warm room so the honey is soft enough to flow smoothly.

The first-year of the Houseboat Honey bonanza required some ingenuity for this part of the process. Since the bathroom is the smallest room in my houseboat, it was easy to crank the heat up, leave the honey there overnight, and then strain it the next day when the honey was warmed up. Sometimes the honey may need to be strained a second time through a finer screen if it is particularly dense with unwanted particles. By extracting the honey in the kitchen and then straining the honey in the bathroom, I was able to make a complete sticky mess out of two rooms in my houseboat.

I then migrated to the living room and made a sticky mess there too, doing the bottling. Why not mess up the entire house? Of the three steps, the bottling requires the most skilful eye, for you must stop pouring at the exact right time, precisely within a quarter of an inch of each bottle's brim. Miss that all-important marker by one-tenth of a second and watch honey overflow all over your hands, seep onto your shoes, and then spread out across the floor.

For me, the best part of processing honey was buying the jars, because the jars come in all different shapes and sizes. As I learned in the advertising business, packaging is 90 percent of the product. Some jars have old-fashioned raised glass relief designs, but I prefer the skinny, tubular jars. They make half a pound of honey look like three-quarters of a pound. You can also choose the plastic squeeze bears and relive your childhood by listening to an audiobook of *The House at Pooh Corner*.

After your honey is safely deposited in the desired jar, you must tightly secure the lid and stick on a label. I wanted my labels to stand out, so I used the same goofy artwork I painted on my beehive boxes once I took ownership of them from Miriam and Len.

The process of making honey, the world's greatest processed food, requires time, patience, and hard work for bee and human. The bees do a three-season nectar-laden dance with Mother Nature, swallowing up summer's sweet floral offerings, then puking them up and fanning them to perfection. The human heist follows on the heels of their labours, just before winter. Unbeknownst to the bees, however, there is a bit of revenge as we Jesse James "wannabees" make a giant sticky mess of every room in our houses. There is justice as well. As responsible beekeepers, we must be sure to leave our girls enough of a store to make it through the winter and replace what we have taken with a food processed by humans.

Together, bugkind and humankind put a great deal of effort into this delightful product that graces many a pantry across the land. And don't forget: honey is delicious, looks great, lasts forever, and is as messy and sticky to serve as all get-out.

THE OUTYARD

When it comes to practical, handy jobs, I take shortcuts. My attention to detail and general handiwork are just plain sloppy; I have two left hands and get distracted easily. I wouldn't have been a good architect, carpenter, or craftsperson. I ignore the old adage "Measure twice and cut once." Why measure at all when it's quicker just to guess? And what's a level for? Beekeeping requires many meticulous skills I lack, but I get by because Jeannie has them.

Being a bee demands those very same skills. Bees are exacting perfectionists focused on building flawless symmetrical honeycomb. There are no shortcuts when constructing the thousands of six-sided cells comprising a hive. If one of those hexagons is a bit out of whack, the whole colony can collapse. Every time I lift out a frame of bees from a hive to inspect it, I am impressed by the degree of collaboration and construction taking place. Bees know exactly what to do and when to do it. Each one is toiling away at the task at hand, head down, focused on a successful end product. Never do you see a couple of bees off to the side of the hive arguing or disagreeing. Never do you see a bee lounging about.

It comes as no surprise to me then that beehives are composed almost entirely of female worker bees. Women, often, simply pay more attention to detail. Jeannie is an excellent beekeeper—she loves perfect right angles and takes on every task fastidiously. Nowhere was this more evident than when we moved four of her beehives to the outyard.

Let me explain what an outyard is. Successful beekeeping is all about the flow of nectar. In spring and summer, forager bees fly around collecting nectar from flowers and plants, bringing it back to the hive to be processed into honey. But what if the fields and flower beds near the hive have dried up as they do in midsummer? Beekeepers can intervene and move the bees to greener pastures. As long as the bees are moved miles—not inches—away from their original spot, they can adjust and make the transition. If you want chubby bees, take them where the getting is good: the outyard. Moving bees to greener, higher elevations for four to six weeks in late summer gives them a better chance of surviving the winter, and gives you a better chance of getting more honey.

Here's an analogy of what moving beehives to an outyard is all about. Let's say you live in a crowded apartment building in some dust-bowl town and a drought has dried up your food source. Then, for some inexplicable reason, your entire apartment building is lifted up, transported, and plunked down in Las Vegas, right behind one of those casinos with a fantastic all-you-can-eat buffet. Every day you visit the buffet and eat all you can, gorging happily and contentedly. If you are like me, you sneak some extra food out in your pockets and store it away for the future. After a month or two of living in Sin City, you become fat and happy. You have spare food socked away in your apartment suite. The dust bowl is a forgotten hardship. Come fall, your apartment building is once again elevated and plunked down back in your hometown for the winter. No problem. Your pantry is full of delicious, life-sustaining manna. That's the concept of an outyard: it's a place to get fat and happy.

Outyards are usually far away from the city on the side of a mountain or in a field in the middle of nowhere. Jeannie's bee club and a neighbouring club joined forces to create a communal outyard for their members. They got permission to occupy a small plot of land from a forestry company that had logging rights to hundreds of recently harvested acres. The clubs were charged nothing for taking over the relatively minuscule 1,000-square-foot plot on the side of a mountain, 75 miles from the outskirts of town. The outyard was surrounded by miles and miles of fields full of a prolific plant called fireweed. After an area has been logged, it is the first plant to grow back. To bees, fireweed is like steak and lobster with baked potatoes smothered in sour cream. They love it, can't get enough of it, and it fattens them up.

Converting raw land into an outyard takes a lot of work. In this case, it required about half a dozen members from the two bee clubs, armed with shovels and shears, to devote eight hours over a weekend in late spring. They cleared vegetation, levelled the ground so hives could be put in place, and installed an electric fence. The electric fence, which was powered by solar panels, was meant to keep bears from crashing the smorgasbord line and going straight for their favourite dish: honey.

Each member of the bee clubs was invited to transfer up to half a dozen hives to the outyard and leave them there for four to six weeks in late summer. That's when I entered the picture—after the hard work had been done. Or so I had hoped, but boy, was I wrong about that!

On the day before the transfer, Jeannie enlisted my help to prepare her four hives. To emphasize why the hives had to be "prepared" for the trip, I must point out that they would be taken up a very bumpy old logging road at 5:00 AM in the back of a pickup truck. We would then take them off the truck bed, hastily carry them into the outyard, and wrestle them into place. These weren't just ordinary crates; they were high-rise homes to

50,000 very alive, very sensitive insects. Think of how careful you are when you put your dog or cat in a crate, load it into your car, and take it across town. Then multiply that: 50,000 × 4 hives = 200,000 precious pets.

Jeannie was quite clear with me that a robust series of rope webbing and carefully screwed-in wooden supports, along with ratchet tie-down straps and fasteners, would have to be adhered to each of the old-fashioned filing cabinet–style hives. Each hive had to be sturdily reinforced to ensure the safety of her bees. The fact that each hive was four boxes tall made the task of stabilizing each tower more challenging. And don't forget the boxes we were preparing were full of live bees. The buzzing inside while we cinched up the straps and tightened the tie-downs was a bit unnerving to say the least. Standing by Jeannie's side with a hammer in my hand, I was given careful instructions that I ignored, forgot, or was incapable of carrying out, apparently. Truthfully, I was uncertain which end of the hammer to hold.

After 20 minutes of strapping and wrapping ropes, using a drill to drive in screws, cinching ratchets, and banging in nails, I impatiently and stupidly proclaimed, "I'm finished with this one. It's fine. I'll move on to the next one now." My precise and exacting girlfriend vehemently disagreed with me, wanting the hive to be fortified with tighter ropes, more vertical wooden slats, and another fastener. She inspected my shoddy work and pointed out what I had done wrong. I took umbrage and disagreed, and then we did something that bees never do—we argued over who was right, instead of working together in harmony for the common good.

After a short verbal scuffle, I begrudgingly acquiesced and returned to battening down the hatches. Silently following Jeannie's instructions, I couldn't help thinking that Harry Houdini couldn't escape from this crate. The rebellious little voice inside my head stubbornly repeated, "We are putting way too much effort into this!" By the third hive I couldn't stand it. I told Jeannie she was taking this whole move way too seriously and it was taking way too long. So we argued some more, and by the time we started working on the fourth hive, it was clear to me that even if it took all night, I should just follow her instructions and do as I was told.

Finally, after two seemingly endless hours, all four hives were prepared and ready to be loaded onto the back of her dad's old blue Toyota pickup truck. In the process of preparing the hives, I had avoided getting stung but struck one of my fingers with a hammer. There was a slight bruise, but it wasn't as big as the bruise to my ego. When we went to bed that night, four super-secure Fort Knox–like towers containing a couple of hundred thousand somnolent bees rested on the back of the truck bed. The indestructible boxes were so well fortified that, I swear, if the truck had happened to be struck by a nuclear missile the next morning, the bees would have been safe. We set our alarm for 4:00 AM to arise well before the bees in order to move them with minimal disturbance.

As you may recall, it is best to move bees in the dark, cool night air. Not because you don't want to draw attention to yourself, but because that is when the bees are comfortably situated indoors. Obviously, when it gets sunny, they are itching to get out and forage for nectar. It is also dangerous to move a sealed hive in direct sunlight because the bees inside can easily overheat and die. In the hot sun, sealed hives turn into ovens, with 50,000 agitated bees rubbing against each other, creating friction that intensifies the heat. So we met the other members from Jeannie's bee club on the side of the highway in the shadowy, cool pre-dawn. Dressed head to toe in bee suits, we drove our pickups and trailers stacked with hives. I worried that any passing motorists seeing a bunch of people in white costumes driving redneck trucks in the dark might get the mistaken idea that we were a chapter of the Ku Klux Klan.

The first thing I did as the five trucks pulled up was discreetly inspect how each hive had been prepared for the bumpy mountain road ahead. There was a great degree of variance. Some hives were well fortified, but none was as secure as ours. Other hives had only one strap or bungee cord wrapped around them. I stopped myself just short of opening my big mouth and saying, "Look, I was right! All we had to do was fasten these boxes together with a single bungee cord. We could have gotten to bed an hour earlier last night." I had created enough acrimony with my lack of cooperation and sloppy shortcuts; besides, the jury was still out on each beekeeper's hive prep and whether or not the hives would withstand the roads ahead.

Six trucks were scheduled to make the trip to the outyard that morning. A guy named Bill from the other bee club, who had organized the whole crazy scheme, drove the lead truck. While we were waiting for everyone to arrive, Bill sheepishly admitted to us that his wife didn't really want him to keep bees. She was worried about him getting stung, which was ironic considering he also trained German shepherds for the police department. After waiting 15 precious minutes for the last truck, we had no choice but to leave without it. It was already getting warmer as the sun rose, and soon the bees would be baking. Each truck was loaded with about six hives, adding up to hundreds of thousands of increasingly uncomfortable and agitated bees per pickup.

The entire convoy, collectively carrying about 1.5 million bees, pulled out and proceeded down the road in the direction of the fresh subalpine meadows and the delicious fireweed plants. Jeannie and I were in the second truck, following Bill closely and worrying about the effect the rising sun would have on our hives. We were already behind schedule, and the sun would be up in 45 minutes. It was a race against time. As we drove, the occasional errant bee escaped from one particular hive on Bill's truck and was briefly illuminated in our headlights.

Soon the first bee struck our windshield. Splat. At 40 miles per hour, it sadly became immediate collateral damage. By the time we had travelled 20 miles down the highway, our windshield was covered with sticky dead bees. Then we slowed down and exited the paved highway onto a rural road that led to a pothole-filled arterial logging road lined with thick fir trees. It got bumpier and bumpier as we slowed our pace to ascend the hill, climbing closer and closer to the restricted licensed forest area blocked by a locked gate.

Bill had made arrangements with the forestry company weeks before to get an extra key cut for the gate. Two other groups of beekeepers had also been granted access to the area and the key would be shared among us. To simplify things (or to complicate them, as I would later learn), the key was hidden in a secret spot under a rock near the gate.

About 10 minutes into the bone-jarring section of road, we noticed more bees leaking out of the one poorly packed hive on Bill's truck. At first it was a couple dozen bees escaping, then a hundred, and the next thing we knew, hundreds and hundreds of bees were flowing out in all directions. His hive had not been properly prepared for transport, a lid or box had been displaced or damaged, and the bees were now madly flying off to who knows where.

Measurements of activities compared to units of time are useful scientific tools. They can also be handy devices in storytelling. For instance, when I describe miles per hour (mph) or feet per second (fps), you know exactly what I am talking about. At this juncture, I offer a new measurement for which I waive all copyright, thus generously donating the term to the beekeeping community: bees per minute, or bpm. It measures the number of bees that pass through any given point and onto a windshield within 60 seconds.

The number of bees escaping Bill's truck right after the five-vehicle convoy left the side of the highway was about 1 bpm. As he increased his speed on the smooth highway surface, the bpm increased as well. While Jeannie drove, I tried to count the bees leaking out of Bill's hive. They were not hard to spot in the headlights of Jeannie's dad's truck, and I added those to the casualties mounting up on our windshield. After about five minutes of driving at 60 mph, I roughly calculated that the bpm had increased to about 5. After we left the paved highway and turned onto the dirt farm road, the bpm doubled to 10. When we hit the even bumpier mountain road, the bpm took a quantum leap to 20 and gradually increased as the sloppily packed hive on the back of Bill's truck slowly disintegrated. Jeannie and I watched in shock and amazement as the bpm hit 25, 30, and then 40. Our windshield was getting all sticky with dead bees.

Bees that leave a hive out in the middle of nowhere, with no way to get back, will die. Bees are communal creatures that can't live on their own. After the bees escaped the hive in the back of the truck, their hive would be placed 20 miles away in the outyard, well beyond their range. If they found another hive to join, one that had been created in nature, the denizens of that hive would likely reject them. As the road got bumpier and the bpm increased, the poorly prepared hive became more morbid. The escalating loss of bee life was tragic.

We honked our horn to let Bill know about the death toll on the back of his truck. He waved at us to indicate he was aware but didn't stop. We realized he was up against time and had to keep his foot on the gas pedal to lead us to the outyard before sunrise, when the bees would begin to sizzle. He had no time to stop and repair the broken hive. He just kept on trucking up the steep hill, and as his speed increased and the road got worse, the bpm hit 100. Bill was an experienced beekeeper;

clearly he was sacrificing that one hive for the greater good of the group.

Finally, he had to stop when our bee transport reached the metal security gate cordoning off the forestry area, which was off limits to the public. It was a heavy steel gate that meant business; stating No Trespassing was unnecessary because without the key to unlock the gate, entering the restricted area was impossible. Bill got out of his truck and searched for the hidden key. While he was looking, Jeannie and I watched in dismay as hundreds more bees broke out of his hive. An angry swarming cloud emerged. It grew and grew, like a dark storm on the horizon. But the sun was rising and there was no time to patch up the poorly prepared, leaky hive. The other million or so bees warming in their hives on the backs of the other trucks had to be released soon or they would all die. It took Bill longer to find the key than anticipated. We fell even further behind schedule. He finally found it, and we passed through the gate.

The last leg of the mountain road leading up to the freshly cleared outyard was the roughest and worst by far. The road was meant for massive heavy-duty logging trucks; the potholes were the size of small refrigerators. Since the area it serviced had already been logged, the road had sat unused and unmaintained for years. It was an almost impassably steep grade. We lost more precious time.

All five beehives on the back of Bill's truck bashed harder and harder into one another and even bounced into the sides of the truck bed with thuds and bangs that I was sure could be heard all the way back in the last truck in the convoy. The poorly packed hive crumbled before our very eyes. We had no choice but to watch its destruction. Then Bill hit the mother of all potholes and the bpm index went through the roof, peaking at about 1,000. So many bees were streaming from that hive on the last mile that it looked like a jet plane's long contrail, and Jeannie had to turn on the windshield wipers. Mayday! Mayday!

We finally reached the outyard at 6:45 AM, under the rapidly warming rays of the morning sun. All five trucks parked hastily in a random grouping amid the lush fireweed. Drivers and beekeepers scrambled out to check their precious live cargo. As we inspected Jeannie's four hives, she didn't need to say a word; her smug grin spoke for itself. Not one bee from her four hives had gotten away. Her attention to detail in packing the hives had saved bee lives. I sheepishly and obediently followed her orders for the next stage of the outyard set-up.

Bill opened the electric outyard fence, and we all hustled to carry our 100-pound hives through and put them into place. All together, we needed to offload about 30 hives from the trucks and transport them 200 yards to the interior of the outyard fence. As it got warmer and warmer, controlled panic set in; each beekeeper rushed to find suitable locations for the hives. Only after the hives were in place could we release the bees from their mobile sun-baked wooden saunas. To make matters worse, the last 10,000 bees in the dilapidated hive on the back of Bill's truck had escaped and were swarming in confusion around us, making it difficult to see. The screened facial veil of the beekeeper's costume also restricts vision. There were so many bees leaking out of that broken hive that dozens of them were now clinging to the screen in front of my face while sweat from the hot sun poured down my brow and ran into my eyes.

Truck gates clanged and hives were transported off the tailgates. Carrying a beehive is a hard, back-breaking, two- or three-person job. The uneven ground, poor vision, and beekeeper collisions made it even harder. Our hives were stacked only four boxes high, so Jeannie and I could manage them on our own. But some of the hives were stacked five or six boxes high, requiring a heavy-duty moving dolly to transport them, except that a dolly does not move so well over a soft, bumpy field. As all 11 beekeepers tussled back and forth between the trucks and the fenced area, we resembled a punk-rock mosh pit. With the 10,000 agitated displaced bees as part of the frantic dance, it felt like we were storming the beaches of Normandy in the Second World War and dodging chaos.

Specific tasks—in order—were required to achieve our beekeeping objectives within the limited time we had at the outyard that morning:

1. Carry each hive from the truck to the outyard.

2. Place the hives on even ground, carefully spaced, with the bee doors facing southeast.

3. Remove the straps, fasteners, wooden reinforcements, and ropes.

4. Last, and most important, remove the duct tape covering the beehive doors and release the bees.

Some of us were still struggling with step one, while other beekeepers were on step four, releasing their bees. Every time that important grey duct tape came off a hive entrance, 10,000 or 20,000 bees burst out of the hot hive and swarmed out into the fireweed for their first breakfast. The number of bees increased from 10,000 to 300,000, circling and dining in an area no larger than a small house. Their buzzing intensified, and the sheer number of them blocked the sun, casting an Armageddon pall over the outyard. It was out of control, and just when I thought it couldn't get any worse, it did.

Remember that sixth truck that never showed up at the meeting place? Well, at that moment, it finally chugged up into the outyard, and as the expression goes, it was "a day late and a dollar short." Whoever had readied the hives in that last truck clearly didn't get the memo. Not only had very little preparation been done to transport the hives, but also they were rotten and falling apart—probably due to wet rot or termites, I wasn't sure which. As the beekeeper attempted to haul each hive out of the back of his truck, the hives simply fell apart. He placed one of them on the ground and the bottom box imploded under the weight of the four boxes on top of it, violently spilling the contents of all five boxes (tens of thousands of bees) into the already insect-laden air.

Two or three beekeepers, including me, rushed to his aid. In my haste to assist, I ran in my bee suit over the bumpy terrain to his truck, but it was almost impossible to see. I tripped over a big dolly and fell face first to the ground. Beekeeper down! I lay there battered, imprisoned in my hot bee suit, and covered in bugs with grains of soil in my mouth. I wished, again, I had taken up stamp collecting.

Pulling myself up, I paused to fully appreciate the number of bees swarming around us. When I looked down I could barely see my feet. Years ago when I went scuba diving, I swam right into a condensed school of fish, and as I frantically paddled to escape, I couldn't see more than a few feet in front of me. The only difference was that the underwater encounter was void of sound, while this situation had such a constant buzzing that it sounded like a couple of vacuum cleaners had been strapped to each side of my head. Regaining my wits, I stood and limped slowly to the aid of Mr. Sloppy Truck;

a few of us helped reassemble the crumbled frames and boxes on the ground.

By this point, all 30 of the other hives were in place inside the electric fence with their front doors wide open. It's no exaggeration to say that about half a million bees buzzed around us, and the newly released bees were not happy campers. They were downright pissed off—hot, in totally unfamiliar territory, and flying around with hundreds of thousands of strange bees from other hives. Show me an unhappy bee, and I'll show you a bee that is about to sting someone.

Despite Jeannie taking precautions before we left her house that morning, duct taping her wrists and ankles and wearing thick clothes underneath the bee suit, she was assaulted. One bee either found its way into Jeannie's suit or was determined enough to plant its stinger through the suit's canvas *and* her inner fleece. Jeannie ran up to let me know she'd been stung. It still astounds me that the one person stung out of our 11-person group was the one person who is allergic to bee stings. The nearest hospital was an hour and a half down the road. We needed to get Jeannie out of there immediately.

On the way up to the outyard that morning, Bill had locked the gate behind him and returned the key to the secret location. I had not paid attention to where that spot was and knew that without the key we would be stuck, so I had to find Bill and get him to explain exactly where the key was. But to find him in the fog of half a million bees—and within the group of people dressed in identical bee suits—was difficult and wasted precious time. Jeannie had sought refuge in the truck's cab, away from the bees, where she peeled off her protective suit in order to inspect the sting on her thigh.

There are everyday directions and then there are important directions. These directions were of the latter variety, and I wished Jeannie could back me up on the listening end. I had already proven the night before that listening isn't my strong point. I find it's best to have two people listen to directions, and when Jeannie and I travel, it is invaluable to have her attentive ears as backup. But as Bill stood in front of the truck talking me through the steps to find the key, Jeannie couldn't open the window and listen because that would have let bees into the truck cab. And I couldn't write the directions down because the bee suit gloves are so cumbersome, it would have been impossible to hold a pen. Besides, I didn't even have a pen. My brain felt like a sieve that couldn't hold anything, let alone potentially life-saving, detailed directions. I tried my best to memorize what he told me and then hopped into the truck, waving the bees out as I opened the door. I started the engine and booted it. Weaving around the other trucks and trailers in the meadow, I manoeuvred onto the bumpy logging road and then proceeded at three times the speed we'd been travelling when we came up.

At all times when tending bees, Jeannie now wisely carries an EpiPen, a syringe loaded with a measured dose of adrenaline, and Benadryl pills, an antihistamine with a side effect of making you sleepy. As we sped down the unmaintained mountain road, she clutched the Benadryl bottle in her hand. Drinking a gulp of cold, leftover coffee from a stainless-steel commuter cup, she downed two pills. Thankfully, her symptoms did not yet warrant the EpiPen.

Still, our tension mounted when, 10 minutes into the drive, we reached the locked gate. As I hurriedly clambered out of the truck, I mentally rehearsed what Bill had said: "The key is hidden right after you see a big rock on the left behind the gate. Look 10 feet toward the fence on the right and there are three little rocks in a row—it's under the middle one." Wandering around the gate in a panic, vagueness crept in and I second-guessed whether the first direction was a right or a left. This treasure hunt was off to a bad start. The sea of rocks alongside the road all looked the same. Which one hid the key? Finally I found the right rock, pounced on it, and grabbed the key. So close to victory, I then couldn't figure out which end of the gate had the padlock. Even when I spotted it, I had a hard time fitting the key into the lock. Jeannie had to get out to help me. I was so happy when she told me then that she was not yet feeling the sting's effect. But we weren't out of the woods yet, figuratively or literally. We had half an hour of a logging road to go before getting back onto the highway. Later Jeannie would confess that as she sat in the truck cab watching me poke the key at the gate's hinge instead of the padlock, she thought we were doomed. Why am I so clueless?

Forgetting to close the gate behind me, I floored the gas pedal and we sped away, fishtailing around corners. Our cellphones were out of range, so there was no one we could call. We were on our own. I had no idea of our fate as we waited for the toxic bee venom to take effect. What a cruel trick for nature to play on us—there we were, transporting these amazing creatures to a wonderful setting where they could eat as much as they wished, and the thanks we got was a bite on the hand (or, in this case, the thigh) that feeds them. Then I thought of the tens of thousands of bees that had been slaughtered because of careless hive preparation. Perhaps it was divine intervention payback time?

Every minute that went by without Jeannie going into shock, I felt more relief. Stings and the way humans react to them are inconsistent and remain a medical mystery. Although there was some swelling around the sting, it seemed to be spreading slowly and not affecting her in any other serious way, whereas her last reaction had resulted in excruciating welts, a quickened heart rate, and hot flashes. Around 10:00 AM, we finally pulled into a strip mall leading into town, where we crawled out of the truck cab, hugged each other, and found a coffee shop. We both drank copious amounts of water and coffee and then returned to Jeannie's place, happy to be alive. The Benadryl's sedating side effect having kicked in, Jeannie slept for the rest of the day. That gave me some quiet time to reflect upon the last 12 hours.

Many lessons in life are learned from mistakes. And that trip to the outyard was full of mistakes. It goes without saying that I learned beehives must be properly fortified and prepared for a major move. If I didn't learn that, I don't deserve to keep bees. The next two takeaways have more to do with the bee club than with me. In hindsight, we could have met at 4:00 AM instead of 5:00 AM. So much stress and tension that morning was caused by a pretty predictable event: the rising of the sun. We could have avoided the intense pressure to get the bees out before they cooked if we had simply left an hour earlier. The final point has to do with releasing the bees once the hives were in place within the outyard. A bit of coordination could have gone a long way. As a group, we should have decided that we would release the bees only once every hive was in place. Finally, Jeannie's allergy to stings remains a concern she is addressing. She has since undergone bee-sting desensitization treatments with an allergist that will reduce her chances of having a reaction in the future.

After reviewing all the dumb human mistakes we made, I considered the bees and the role they played on that calamitous morning when the Keystone Cops of Beekeeping moved them to an outyard. Did those savvy little creatures sense our missteps and collectively take advantage of us? Had the 1.5 million bees just had enough? Were they fed up after the cruel shake-and-bake treatment we put them through on the trucks? Did they consciously decide to retaliate? Were they smart enough to single out the one member of our beekeeping group who was allergic to bees? Did they appoint the strongest bee to attack her?

I decided that, rather than ruminate on negative subjects like "bee revenge," I should focus on the positive. We had successfully transferred 30 hives of bees, with a total population exceeding the population of the city of Phoenix, to where they were safely lodging in prime nectar country. All those bees were now happily chowing down on the bee equivalent of filet mignon: fireweed. I had learned a great deal, much more than I would have sitting at home reading a how-to beekeeping book, and Jeannie and I were still on speaking terms. All in all, the day had been worthwhile. Exhausted, I popped a Benadryl and joined Jeannie in a long afternoon nap.

As a postscript to this story, I need to point out that beekeeping at the outyard is different from tending a single hive in your backyard. Many beginner beekeepers can safely tend those single home hives with little protective clothing. The outyard, however, is so concentrated with bees that it makes the proper protective gear a life-preserving necessity.

GOD SAVE THE QUEEN

The queen bee, as her title implies, is the royal bee all worker bees and drones love, respect, and worship—the exalted empress for whom they work their brown-and-yellow tails off. She is *it*: the monarch with the most. The future of the entire hive rests upon her shoulders, or, more specifically, her thorax. As a beginner beekeeper I had a tough time spotting the queen, an important skill I needed to develop if I wanted to keep on with it.

Given the pivotal and prestigious role the queen bee plays within a colony, she sure doesn't stand out visually inside the hive. To the untrained eye, the queen—nestled deep among tens of thousands of her offspring—has pretty much the same shape, basic colour, and markings as all the other bees. Her only distinguishing trait is that she is a bit bigger. I'd say that all the worker bees are about half an inch long and she is about three-quarters of an inch long, or about 50 percent bigger. Given that a quarter of an inch is quite minuscule, the queen doesn't exactly pop out. Yet, as a beekeeper, when you check the inside frames of your hive, there is nothing more important than ensuring she is there. A queenless hive is eventually a dead hive. When beekeepers open up their hives and check inside, their number one task is to make sure the queen is present, healthy, and laying thousands of eggs every day. If you are ever looking to make conversation with a beekeeper at a party, just ask, "How is your queen doing?" If the response is "I can't even find her," you'll know that person is a beginner like me, or worse, their hive is queenless and will soon be toast.

I learned from my beekeeping sister that you can use a special red felt pen to mark the queen bee so you can see her more easily. A special bug Sharpie! You carefully draw a dot on the queen's back with the pen's bee-safe ink. This sounded convenient, but unfortunately my skills were not developed enough to locate her, adeptly pluck her out with my fingers, and start doodling on her. I'd probably squish her to death, or knock one of her antennae off while accidently painting a dot on her. However, I was intrigued by the concept, and it led my mind to wander away, as it often does, to comparable real-life experiences from my past.

In 2004 I lived in London for a year, working on the city's bid for the 2012 Olympic Summer Games. Given my work as one of the directors of this historic bid, I was invited, along with 75 of my colleagues, to Buckingham Palace to meet Queen Elizabeth. The reception was held in a lavish ballroom. This queen, however, was 50 percent *smaller* than everyone else. Her tiny 100-pound frame made her easy to spot, but it was the diamond tiara on her head that was the dead giveaway. I can only imagine what would have happened if I'd pulled out a red Sharpie and drawn an identifying dot on her turquoise dress. Something tells me that the next yard I visited wouldn't be the outyard but Scotland Yard.

Like the queen in my hive, Queen Elizabeth had attendants fussing over her and serving her during the reception. I distinctly remember the queen radiated a lovely royal perfume scent—a cross between roses and lemons, a refreshing combination of scents unlike any I'd ever smelled. The queen in my own hive also emits a pheromone odour perceived as special by most of the bees in her large, busy colony.

As the queen's entourage hovered close to the tiny woman and saw to her every need, it was clear to me that, just like the queen in my hive, Queen Elizabeth's health, well-being, and longevity were paramount to the survival of the colonies. The future of a bee colony, like the Commonwealth, depends upon an abundance of healthy offspring from the central queen. This was where the queen in the hive on the back of my float home had Liz beaten hands down. Whereas Queen Elizabeth produced only four offspring—an heir and three spares—my tiny little queen would lay over a million eggs in the course of her lifetime.

In my attempt to get up to speed on beekeeping, I had already ascertained that checking the queen's egg-laying patterns occasionally is a good rule of thumb. History shows us that Britain's royal queens have at times been subject to a similar vigilance. However, as a beekeeper checking on Her Royal Highness of the hive, I discovered the hard way that you should never leave checking on her until the day before you are about to go out of town.

I'm sure you know what it is like when you are about to go on holiday—you rush around your house madly, ensuring that everything is okay before leaving. If you are like me, you usually put off the preparations until the night before you go. Such was the case in early May when Jeannie and I were about to depart for a three-week cycling trip from Vancouver to Saskatchewan. After ensuring that the propane on my float home was turned off, the water connection was off, everything was unplugged, and the heat was turned down, I went to check on the bees. Bad idea. Never check your bees when you are about to leave, because if you discover something is wrong, there is likely nothing you can do about it.

I'll give you three guesses who wasn't at home when I scrutinized every frame in my hive. You guessed it: Queenie. But you may also be thinking, "Didn't he just admit he is useless at spotting the queen? So how did he even know she was missing?" You are right; I *am* lousy at spotting the queen bee. But a hive without a queen displays some definite behaviours that alert even beginner beekeepers of her absence.

When the queen vacates the colony, her pheromone smell leaves with her. With her reassuring scent gone, the leaderless colony becomes stressed. Since minuscule amounts of her pheromone perfume are usually left behind, the worker bees immediately start fanning their wings like crazy to distribute whatever essential regulating chemicals are left. Therefore, a queenless hive often presents a louder, more distinctive buzzing than a hive with a queen. It is important to listen to the hive before you open the lid and listen for an elevated buzzing sound.

The next sign you have a queenless hive makes even more sense. Without the queen, no new eggs will be laid. So, as you look through each 7-by-19-inch frame, you need to peer carefully into the thousands of tiny six-sided cells to see if there are signs of eggs, larvae, or pupae. I don't want to get too technical here, but since this is a book on beekeeping, it is important to note that three days after a queen lays a fertilized worker bee egg, it will turn into a white, creamy larva in the cell. After seven days, it forms into a pupa, which is like a bee embryo. After 21 days, it hatches into a fully functioning fuzzy little bee. What I just shared is one of the fundamental beekeeping principles.

If you can't spot any of the three egg/brood stages, it means the queen has not been there for 21 days. As I inspected my hive the day before my trip, I was delighted that many of the tiny cells were full of delicious Houseboat Honey, but the rest of the cells were void of eggs at any stage of development. My hive had a healthy number of bees, all building comb and packing away honey for a future they didn't have. With Big Mama gone, the hive was destined to die. In order to emphasize this point, it is important to understand that the average worker bee lives for only six weeks in the summer. Worker bees, the vast majority of the bees in the hive, have an extremely short lifespan; they work themselves to an early death. The queen, however, lives for a few years. It is her job to continually replenish the stock of offspring.

On that Sunday night before leaving on holiday, I was pretty certain I had trouble on my hands. Was I signing the death certificates of thousands of bees by doing nothing and leaving? The obvious solution was to purchase a new queen and plunk her in the hive. But where do you purchase a queen bee at 10:00 PM on a Sunday? I turned to the beekeepers' best information source: the Internet. I quickly learned that although some people in the Vancouver area do breed queen bees, they are rare and hard to get a hold of late on a Sunday night. So, as I walked up the ramp off the dock early the next morning to leave on our trip, I said goodbye to the girls with a heavy heart, knowing I might never see them again.

As we pedalled from Vancouver west through two provinces of vast bucolic farms on our way to Saskatchewan, we passed dozens of beehives placed next to different crops, open fields, and treed meadows. Cycling with such enthusiastic beekeepers as Jeannie, Miriam, and Len meant that we just had to stop to have a look each time we came upon some hives. It did cross my mind, given the isolated roads and the total lack of people, to pinch a queen from one of those hives. But I came to the conclusion that ending up in a jail cell in Maple Creek, Saskatchewan, because of a bum bee rap wasn't a good way to begin my new hobby or end my holiday. Besides, how would I get a stolen queen back to Vancouver? My bike's luggage panniers were full. I had no choice but to pedal onward and hope for the best for my hive back home.

When we finally returned to Vancouver and I had lugged my gear down the boat ramp, the first thing I did was hustle out to the back deck to check on my hive. I anxiously scrambled into my white bee suit, adjusting the zippers around my neck to secure the head veil and hat, and braced myself for the worst. Given the short life cycle of the honey bee, I knew thousands of bees would have died of old age since we left, and given the fact that no new bees would have been born to bring up the rear, I expected the hive to look a lot different. I expected it to be a whole lot emptier.

And it was, but not shockingly so. I'd say my hive had about 30 percent fewer bees than it did three weeks earlier. The bees seemed a bit lethargic, but the hive was still very much alive with even more honey stored in it.

There is a third sign that your hive has no queen, but it takes a few weeks to discover. As I studied each frame, I noted hundreds of large brown cells, crusty and overflowing with a dark brown cap. When a hive has been queenless for a few weeks, some of the unmated worker bees begin laying infertile eggs. They kind of freak out and start laying useless dud eggs that will hatch into male drone bees. A small number of drone cells can be found in a healthy hive, but my hive was becoming overrun by them, confirming I needed to get cracking and find a new queen.

I got out of my bee suit and went inside to find the contact information for a local beekeeper who, I had heard, raised queens. His name was Arnold, and I chuckled to myself when I called him and said, "Hi, my name is Dave, and I need to buy a queen." It may have been funny to me, but it was regular business for him. He raised queens for cash and had one that I could pick up in two or three days. When I told him I couldn't wait that long, he agreed to meet me within half an hour and sell me one for $50. Wanting to get that egg-laying goddess installed and producing offspring as quickly as possible, I sped across town to where Arnold was waiting for me out on the street in front of his house. As I handed this stranger two crumpled-up 20s and a 10, he discreetly reached into his shirt pocket and pulled out a small plastic cage about the size of a Bic cigarette lighter. It felt like a drug deal. I asked him if she was mated, from good stock, and if he raised her himself, just like you might ask a marijuana dealer about the potency and origin of a bag of pot. Anyone watching us out there on the street wouldn't have been blamed for calling Crime Stoppers.

Arnold advised me to keep the queen in my pants pocket on the way home because the warmth of my body would be beneficial to her well-being as I transported her. I was a bit nervous placing a large stinging insect so close to my private parts. I double-checked to make sure the lid of the small plastic cage was tightly fastened. I wanted to ask Arnold how he raised queens, but we were both in a bit of a hurry, so I hastened off with what I hoped was the fertile solution to my hive's problems.

The next stop was my sister's place to pull three frames of bees out of her healthy hives. If the new queen nestled warmly down there in my pants was going to lay eggs, she needed young nurse bees to nurture the baby bees after they were born. Nurse bees feed the baby bees, clean their cells, and generally tend to all their needs (as the name would imply) for the first week of their lives. The role of the nurse bee is one of the first jobs in a young bee's life. Bees go through several work phases as they mature, with the last assignment being outdoor forager. All of the bees in my geriatric hive were either older foragers or drones, which do nothing but eat. The rough-and-tumble forager bees will have nothing to do with baby bottles and pollen pablum.

The three frames of Miriam's young nurse bees and brood would give my future queen's eggs a fighting chance. My sister's healthy hive had thousands of bee eggs, or "brood" as they are called, and thousands of young, healthy nurse bees. Mixing and matching bee frames is a common beekeeping practice, I later learned, and it is always good to have someone in your family or a friend nearby with a healthy hive. Speeding over to Miriam's, I knew one thing: no matter how much the three frames of bees I was getting from my sister's hive needed warmth, I wasn't going to put them in my pants. Good thing I had a special cardboard box in my trunk called a "nuc box," which I would place the frames into for transport. It took about an hour to go through all of Miriam's frames and find three with just the right mixture of nurse bees and brood. Plus I had to be extra careful not to accidentally steal a queen from her hive.

After leaving Miriam's place on that June afternoon with bees on my mind, bees in my trunk, and a bee in my pants, I was distracted. I felt a real urgency driving speedily back to my own hive. I ran through a yellow light and got a ticket. As every hour went by, a dozen or so bees in my hive on the float-home deck were dying of old age. I thought of a depressing new bee acronym: dead bees per hour (dbph). Meanwhile, the new queen was doing absolutely no good sitting in my pocket. I had to get her back to the hive to do some old-fashioned egg laying ASAP. Plus, I had to introduce those three new frames full of bees and eggs to my existing hive. On the way home I tried to recall if I had an old newspaper lying around. I'd need a newspaper for the next step in the transfer process. When you introduce a large number of new bees to an existing hive, as I was about to do, their immediate instinct is to kill the unfamiliar queen. Bees can be real pricks.

I have to admit here that I like bees the most when there are only a few of them on a flower stem. On a carefree, sunny summer day, I could watch a bee for 20 minutes, observing as it crawls all over a flower's stamen and navigates the petals and stigma in search of nectar. When just a few of them are flying around, it is interesting to see them playfully interact as they sample the flora. Bees are cute and fuzzy in small numbers. The minute you open a hive, though, they overflow like a volcano—a pulsating, buzzing, confusing, bloodthirsty, single organism that, quite frankly, is a bit gross and definitely scary. They are still interesting, but they lose their charm, cuteness, and innocence when they morph into a mass of 50,000. A bee on its own conjures up childhood memories of Winnie-the-Pooh and silly cartoon bugs. The inside of a beehive is more like a Stephen King horror novel. It's creepy watching them crawl all over each other. It gets weirder when you realize that, although most people are okay with one or two stings, if all the bees in the hive decided to sting you at the same time, it would probably kill you. And they are not very welcoming to each other when it comes to hosting new bees. Foreign queen bees smell differently than they do, so they kill them.

Back to the newspaper. To combine a hive with frames of bees from another hive, you start by placing a piece of newspaper, with a few slits in it, on the top box of the existing hive. Then you put another box with the new frames of bees and brood in it on top of that paper. Normally the boxes have no barriers between them and the bees can flow freely. With this method, the thin layer of newspaper will contain them in their respective boxes and eliminate fighting. After a day or two, the two boxes full of agitated and antagonistic bees will settle down. They get used to smelling each other and to the idea of new neighbours moving in upstairs. Their scent travels through the slits in the newspaper immediately, and eventually the bees do too.

After a day or two, the bees on either side of the paper start to nibble through it, and slowly they migrate both upward and downward. The beauty is that they now travel in peace because they are used to one another's smells. Hard to believe that a thin piece of paper can have such an amazing effect. I leafed through the latest *Vancouver Sun* trying to decide which page to use; I knew it didn't really matter, but I chose one from section "B."

So there I was on the back deck of my float home, where I had just introduced two warring tribes of bees, each tribe in its separate box determined to fend off any new smells. Talk about a harsh anti-immigration policy. Their coexistence and future literally relied on a paper-thin détente.

With the newspaper sheet delicately in place, I was ready for step two in this crazy melting-pot exercise. The new queen was now ready to join the hive and take her rightful place reigning over her subjects: her brood and her nurse bees, her foragers and drones. With plenty of fresh brood and nurse bees from my sister's frames, I was growing optimistic that this whole half-baked scheme of mixing and matching hives and a new queen might just work.

Not so fast. The same tendency the rank-and-file bees in my hive had of killing a new queen with an unusual new smell also applied to the way Miriam's nurse bees would welcome or reject a new queen. The lemony-rose queen smell I experienced at Buckingham Palace didn't bother me in the least or make me want to harm Her Majesty, but the bees in both the existing colony and the three new frames needed time to adapt to this new queen's scent. Otherwise, it would be lights out for Queenie. I told you that certain aspects of beekeeping are straight out of a horror novel. The new queen and her distinctive royal odour had to be eased slowly into her new surroundings. When Arnold handed me the plastic queen cage earlier that day, he explained that after I popped the top off, there would be a layer of condensed, sugary cotton candy blocking the entrance, thick enough to take a couple of days for the worker bees to chew through. And chew through it they would. Remember how bees love anything sweet. The plastic chamber is designed so that by the time the bees chew their way through to the new queen, they are used to her smell and will treat her like royalty for the rest of her life, instead of killing her. *Sweet.*

Beekeeping goes from moments of excitement and drama to long periods of nothing happening. It is a lot like baseball. With the new queen and new brood in the hive getting used to one another's smells and chewing through the newspaper and the cotton candy blocker, all I could do was wait for about five or six days before going back in to see if I could spot any new eggs. It was tough to resist opening the hive and poking around, but doing that is disruptive and only pisses the girls off. Patience is not a strong suit of mine. The five days of waiting to see if the new queen had "taken" passed very slowly.

When it was finally time to open the hive, the only good news was that the newspaper had almost completely disappeared and the two tribes, having eaten through it, had merged peacefully into one hive unified by a new smell: the rich fragrance of solidarity. The bad news was that if the newspaper had still been there, its headline would have read: QUEENLESS AND EGGLESS HIVE LEADS TO BEGINNER BEEKEEPER BEING CHARGED WITH NEGLIGENCE. Don't ask me what happened. Maybe the queen was hiding. Maybe my bees couldn't accept her smell and killed her. Maybe a wasp got in there and murdered her. Maybe she got injured in my pants. Maybe she just flew off somewhere else. All I know is that after half an hour of searching as hard as I could, I couldn't see a queen anywhere in the hive. I didn't see a single new egg either. Jeannie and I agreed it was possible the new queen was in there and I had just missed her, although there should have been signs of egg laying. We decided to give her another week to make sure she wasn't in there hiding and about to go on a delayed egg-laying spree. *Uh-oh*. More patience required.

Seven more extremely long days passed before I popped off the hive lid. What I saw then was nothing but lots of honey and a few newly hatched bees from the brood I had borrowed. In terms of fresh-laid eggs from the queen, I couldn't find one, not one measly egg. I knew it took 21 days for an egg to hatch, so I knew that the baby bees in there were not from my queen. She was dead, gone, or infertile, but one thing was for sure: she wasn't producing. I felt like the $50 I had given Arnold had grown two wings and flown off with the missing queen. I was back to square one. But, looking on the bright side, at least I was learning some of the beekeeping basics.

I was on two months of borrowed time now, with not one egg laid in that entire period. The hive was on its last legs and would most certainly die soon. I had been told that beekeeping was fun, but it was turning out to be a stressful deathwatch. The old hackneyed expression "If at first you don't succeed, try, try again" is especially fitting for new beekeepers like me. I wasn't about to give up. Good thing I had a plan "bee." I'd get another new queen for the hive; however, this time I'd go to a bona fide queen supplier. A morning Google search led me to a bee supply store in Richmond, a nearby suburb, which I felt would be a lot more legit than meeting some stranger on a back street.

Bob, the old gentleman who owned the place, told me over the phone he had just gotten a fresh supply of really good queens from Hawaii, and he would sell me one for $45. It is common for bee suppliers around the world to get their stock from tropical locations. He was open until 5:00 PM, the price was right, and I've always liked hula skirts and ukulele music. So I hopped in my van and bolted over to BC Bee Supplies. Bob's queen cage was a bit different from Arnold's plastic one; it was wooden with a fine metal mesh screen on the front of it. And the queen wasn't alone; she had half a dozen Hawaiian attendants in there with her, feeding her pineapple pollen or macadamia nut nectar, I suppose. This time I knew it wasn't as simple as plopping her in the hive and having eggs the next day. Even if the queen survived in my hive, it was too sparse on newborn nurse bees; I needed to introduce more. So, on top of my order of a new queen, I asked for "a side of brood." Then, just like in some greasy spoon diner, I barked out, "To go, please."

This time I got only one frame of brood from Bob, with no bees—only the capped brood. After pulling the frame out of one of his hives, Bob used a soft kitchen broom to gently brush off the bees. One frame with no live bees was all that he was willing to sell me, and I was okay with that since some of Miriam's three frames of brood were still in my hive. It was almost like déjà vu driving back to the float home, a bit more carefully this time, carrying a caged bee about to be plunged into the darkness of the hive box, along with the frame full of egg brood in my trunk, which would ideally hatch in time to nurse the eggs I hoped she would lay. I reflected that although we humans perceive nature to be simple, we tend to complicate things when we intervene. All I wanted was a few jars of damned honey. It would have been a lot easier to just go to the supermarket.

When I got back to the river, I settled the new frame into my hive. Since it had only brood eggs and no bees on it, a newspaper wall was not necessary. However, the entire hive had to get used to the tropical suntan lotion smells of the new Hawaiian queen and her attendants, so, like last time, I placed the small, closed wooden cage in between a couple of frames and left it there for two days. When the time was up, I slowly pulled the cork out of the top of the wooden box, exposing the cotton candy to the escape hatch. I chuckled when I saw the stencilled black letters on the back of the box indicating she was a Kona Queen. If this worked out like I planned, she would become a River Queen, and if she was going to stick around like I hoped she would, it was going to get a hell of a lot colder in the winter up here in Canada than it did in Hawaii. After again tucking the queen in her cage safely between two frames, I put the lid back on top to seal the hive, and the waiting game began. *Again*.

A few days later, I went back into the hive to check, and the cage was empty. Good start.

So the new queen bee was in the colony, but as Shakespeare would have said way back in the 1500s when Elizabeth I was doing her best to produce offspring: To lay or not to lay—that is the question. The answer would take another week. In another week, when I lifted the box lid, there they were: eggs, glorious eggs! Row upon row of tiny white diamonds with a slippery, wet sheen, each one bursting to the seams with endless potential—the bees of tomorrow.

I knew by their appearance the new queen had just laid them in the last day or two. With a healthy queen in my hive laying thousands of eggs a day, I could finally relax. The journey from being queenless and clueless to reviving my hive had taken a few months, but there was a real sense of satisfaction knowing I had created the right environment for the new queen to do what she was meant to do, which was to lay a million eggs over the next few years, each single egg a tiny white, creamy, perfect little grain of life.

Standing in front of my hive, holding up the frame so the sunlight could shine into the back of each six-sided cell, I marvelled that in another day or two the eggs would slightly quiver and become little worm-like larvae. Then the larvae would lounge and incubate in their cells and eat and eat and eat—honey and pollen and royal jelly and anything else the young nurse bees would feed them. Bees need the same honey we love to put on toast and in tea, to sustain life, to grow from egg to bee. Nine days later they would be 1,000 times the weight of the original eggs. About three days after that, the worker bees would seal the tops of the cells. That's when the enclosed creatures enter the pupa phase of development. They do not eat or drink until day 21, when they emerge as perfect brown-and-yellow bees.

By now you know what their first task in the hive will be: cleaning the cell they emerged from. Then, after a couple of days, they become nurse bees. The circle of life keeps on spinning with Mother Nature taking over. Despite my clumsy, incompetent beekeeping skills, I had set up my new queen to succeed. By ensuring eggs would keep flowing daily out of the back end of my precious Royal Hawaiian Highness and were cared for, we—Queenie, Mother Nature, the nurse bees, and I—saved the hive.

After learning how important the queen is to the hive and how hard she works, I decided not to indignantly mark her with a red felt pen. Instead I gave her a tiny tropical fruit-motif tiara, a miniature version of Queen Elizabeth's, studded with microscopic diamonds, rubies, and sapphires, which is far more fitting for a regal, productive, incomparable insect of her stature.

THE MIGHTY MITE

I arrived home one day and found a handwritten note from one of my cleaning ladies: "You are out of dishwashing soap, also we found 1,274,867,325,012 ants in the house. The plastic bag that has the beekeeping equipment in it is teeming with them."

She was right: the ever-present ground enemy of the honeybee had infiltrated my float home. It was my fault of course; I had left some empty hive frames and small tools in an open plastic bag in the spare bedroom.

My bees and I have this in common: neither of us wants ants in our humble abodes. Ants are attracted to sweets like sugar and chocolate, and when they visit the float home they go straight for my kitchen counter and the honey jar. I have come home in the past and caught half a dozen of the irritating little black arthropods adhered to the jar's loosely attached lid, pigging out or gobbling drops of honey oozing out of the sticky hole on top of the plastic honey-bear's head.

In the great outdoors, ants go straight for the honey source too: the beehive. Ants aren't the bees' only problem though. Bees get attacked not only by ants on the ground but also by predators in the sky, such as wasps, bees from other hives, and airborne diseases. Then there are the tiny parasites called mites. It's a wonder the poor bees can make honey with all the time they spend keeping would-be attackers at bay and dodging microbes.

Wasps and bees may look similar but they are completely different, as different as cats are from dogs. Wasps eat meat. I'll leave it at that for now and let you arrive at your own conclusions about which insect you would prefer to spend some quality time with. Wasps eat honey too, but of course they are incapable of making it themselves. So, just like ants, they try to illegally enter the hive to steal honey. In law-enforcement terms this is called a Bee and E.

With armies of ants, legions of wasps, and foreign bees attacking the sweet honey stash, what's a beehive to do? Well, just like at Buckingham Palace, beehives have guards stationed at the entrance. You already know that shortly after bees are born and clean their cell, their first task is to become nurse bees. Being a nurse bee is a slack and safe job—all they have to do is scrub cells and crawl around, ensuring all the baby bees have enough honey to eat. After only two weeks on that job, the nurse bees develop flight muscles and stinging glands and become capable of defending the hive. They are then promoted to guard-bee status. Healthy hives have up to 100 guard bees on duty during peaceful times and can recruit thousands when a threat occurs. It's pretty straightforward. The guard bees position themselves at the hive's door and give the sniff test to any intruder. No passport or credentials are required for entry into the hive; the pheromones the insects emit are identification enough.

As beekeepers, we can assist guard bees by narrowing the size of the wooden entrance through which bees access the hive. We use a special piece of wood called an entrance reducer, which is typically 15 inches long and half an inch high. It has an assortment of different-sized openings or holes to choose from. If we observe an inordinate number of suspicious-looking characters hanging around, we can reduce the size of the hive's entrance down to 2 or 3 inches wide, allowing the guard bees to be more effective and threatening. It's like a bouncer at a nightclub keeping out the rowdies—their job is easier if they're guarding a 3-foot cubbyhole rather than a 30-foot barn door. Bees, like nightclub bouncers, challenge aggressive antagonists, but bees will fight it out with a ne'er-do-well wasp or ant until one of them drops dead. If you ever get a chance to stand near a hive and observe the entrance, it's pretty fascinating.

Wasps, ants, and bees from other colonies are easy to spot. I have learned that robber bees from other colonies are often a slightly different size and colour, and they may display more aggressive behaviour than the bees in my hive. While these intruders can be controlled to a certain degree by adjusting the hive's entrance size, it's the pests you can't spot that cause the real havoc. Enter the ferocious, mighty, yet nearly microscopically tiny varroa destructor mite.

Any parasite that has the word *destructor* in its name is not a welcome guest in my hive. These little villains threaten beehives all over the world; they are universally nasty. Interestingly though, the varroa mite was not seen in Europe until the early 1970s and didn't make it here to North America until 1987. Since then, hives on this continent have been dying or "collapsing" at an alarming rate. It may be that modern beekeeping practices, which often position many hives in close proximity to one another, are providing prime conditions for the varroa to thrive. These big hives are then trucked en masse all over to pollinate various monoculture crops, thus helping varroa mites migrate as unwanted stowaways. If you talk to any beekeeper who has been around for a while, you'll hear that the infestation of this strain of mite kills off more hives every year.

It gets worse. The little creeps are bloodsuckers. Technically, though, bees don't have blood; they have something similar called hemolymph. But you get the idea: varroa mites are mini-vampires that literally suck the life out of your bees and thus your hive.

There are books, research papers, and lengthy Wikipedia entries written on the varroa destructor mite, all of them hard for the average person to understand and a tad on the dull side unless you have your master's degree in entomology. But it's really pretty simple. Varroa destructor mites are black and about the size of a coarsely ground grain of black pepper. They make it past the guard bees and into the hive through ingenious, sneaky manoeuvres. A pregnant female varroa might, for example, attach herself to an innocent worker bee out there in the field minding her own bee business, foraging around the pretty flowers. When that worker bee comes back to the colony, she passes the sniff test from the guard bees because the stowaway varroa is odourless, and the tiny freeloading mite enters the hive undetected. Then that pesky pregnant parasite hops off the innocent bee's back and starts looking around the hive for a cell with a bee larva that is ready to be capped by a nurse bee. Once the varroa crawls into the larva's cell, the tiny mite is happy in the private, capped environment because she can feed on larva protein and begin laying her own microscopic eggs in a perfect incubating chamber.

The eggs the varroa lays in comfy bee-larva land are typically one male egg and a few female eggs. The mite eggs hatch in the capped cell and then the male mite mates with his female siblings; they like to keep it in the family. Next thing you know, the tiny cell the varroa has infiltrated is packed with more malicious mini-mites. This is big trouble because the mites could eventually take over the entire hive.

After the newborn mites leave the cell, they have free rein to wander throughout the entire hive, and all hell breaks loose. This is where watching those old vampire movies will give you a leg up on beekeeping knowledge. These parasites live by sucking the life out of the worker bees. I have actually put one of my honeybees under a powerful magnifying glass and spotted two mites on the poor girl.

Life in the hive is highly organized, and the bees have a ton of work to do. Between distributing pollen, cleaning up after one another, raising babies, filling cells with honey, defending the entrance, and dozens of other jobs, each bee is, as the expression goes, busy as a bee—until the varroa mite sucks the energy out of them, that is. A deliverer of deadly danger, that evil arachnid can quickly wipe out an entire bee colony. There are several ways to eradicate the mighty mite. A cross at the hive's entrance and a clove of garlic are not among them. I'll address the two most popular, one gentle and one horrid.

Icing sugar has a consistency kind of like chalk dust. If you gently go through your hive and sprinkle icing sugar all over your mite-infested bees, they get covered in the stuff and then preen each other to get it off. While they are using their tiny bug arms and hands to rub, clean, and scratch each other free, they also manage to knock off many of the varroa destructors. The mites fall down to the bottom of the hive through a wire mesh screen and onto a white plastic collector board, where they eventually starve to death. The upside for the bees is that icing sugar residue is food for them, and after their preening session, they get to eat whatever icing sugar fell off. This method is interesting and safe, but not as effective as the nasty, heartless one: vaporizing.

Oxalic acid is a poisonous industrial chemical used for bleaching pulpwood and cleaning rust, as well as various uses in the laundry and textile industries. It is also used to seal marble statues and to remove magnesium and iron deposits from quartz crystals. When half a thimbleful of the salt-like white oxalic grains are heated up to a super-hot temperature, the process rapidly creates a toxic vapour. For some strange reason, honeybees are not affected by it. However, it is not recommended that beekeepers breathe it in. I have had a whiff go up my nostrils, and it is harsh. As for mites, it kills them instantly. This process is called vaporizing your bees. Beekeepers fall into two camps: those who vaporize and those who don't. It can be a touchy topic, but the sad truth is beekeepers who don't vaporize often see their hives die off.

The guy who invented a gizmo to vaporize bees belongs to Jeannie's bee club, and he sells them all over the world. It is a flat, grey, cast-iron plate about the size of a deck of cards with a heating coil attached to it next to a small pocket that holds the acid. And get this: a car battery powers the whole ingenious contraption. First, you pop the hood of your car, drag out the battery, and haul it over to the hive. Then you place the vaporizer under the hive, hook up a couple of wires to the positive and negative terminals of the battery, and let it sizzle under the red-hot electrical current for one minute. It is kind of like jump-starting your hive. After filling the hive up with the deadly gas, you leave it sealed for 10 more minutes. Then, holding your breath as you come back, you remove the vaporizer. To keep the mites at bay, you must repeat this procedure every few months or when needed. And a tip for new beekeepers: don't forget to put the battery back in your car.

Years ago when Miriam was a beginner beekeeper like me, she treated her bees for mites. The next day, not thinking it through, she harvested a few dozen pounds of honey. A week later, she jarred the honey and gave it away to grateful friends. Lying in bed one night, she remembered reading a warning that after treating your bees for mites, you should wait two weeks before pulling honey from the hive. Tossing and turning on her mattress, she feared the worst that long, sleepless night. Fortunately, most of her friends had not opened the jars, and the ones who had, lived. Their constitutions must have been stronger than that of the average mite.

The seemingly innocuous honeybee has lots of enemies. In addition to the mighty mite, which is a significant threat to bee colonies worldwide, there are spores, fungi, and other sorts of micro-organisms that if left unchecked and untreated will destroy your hive. I call these the "mean microscopic maladies." They are really disgusting and have names that you have probably never heard before like nosema, chalkbrood, and the dreaded American foulbrood. There are noxious chemicals to treat them too.

American foulbrood is the most serious; it is caused by a spore-producing bacterium. About 2 percent of hives in North America will get this nasty disease, which has no cure or fix. You have to burn all of the wooden frames, inside covers, and the bottom board of an infected hive. Then you have to get a torch out and scorch the inside of your boxes to rid them of all foulbrood spores. These bacterium-ridden spores mean business, and nothing short of fire will stop them. Not only does American foulbrood wipe out your entire bee colony, but you have to replace most of your equipment too.

Thinking about all these different insects, parasites, and microbes, something struck me as ironic. Most hobby beekeepers get into it for the love of nature, to do something good for the environment. We love our honeybees and work hard to protect them. But wasps, ants, mites, and slime—well, they are part of nature too, yet we do our best to wipe them out. It feels a bit hypocritical. I wonder if, by interfering with nature to help our bees, we're really doing as much good as we think. Maybe by using chemicals to help our bees thrive, we will throw off the mite population, which will throw something else out of whack, which will come back to bite us, literally. One thing that for sure never changes is human nature. As long as bees keep my honey jar full, I'll watch their backs. Too bad for the mites. They haven't figured out a way to make something I'd find as valuable as honey; otherwise I'd be pampering and protecting them too. But for the time being, when I come back in my next life, I hope I come back as a honeybee and not as a varroa mite.

SHOW ME THE HONEY

Most how-to beekeeping books are written by well-meaning zealots who are so enthusiastic about their labour of love that they invariably leave out some of the less desirable parts. At this point, you might actually be considering keeping bees yourself, and if so, there are a few practical things you should know. You already know you'll get stung repeatedly, there's intense hazards in transporting hives, you will have no opportunity to travel, and invaders and parasites are par for the course. However, you may still be considering doing it yourself. Now, I don't want to be a Doroghy Downer, but if you want to keep bees, get out your wallet.

Don't get me wrong; beekeeping is a great hobby. Simply stated, though, there is no money in honey. How many rich people do you know who made their fortune in honey? Oil, gold, stocks, or real estate, yes. But honey? No. Most amateur beekeepers I have met have either low or no expectations about a return on their investment. They have a toned-down, realistic attitude. If you ask them to describe their hobby, they'll probably say it's interesting, rewarding, and good for the planet. Being a self-proclaimed capitalist apiarist, it's important to me to get some kind of return on my investment—show me the honey!

I was so happy that first year when we harvested 100 pounds of award-winning Houseboat Honey. But I really had very little to do with the process; Miriam and Len did all the heavy lifting. After that promising start, I was skunked with little or no honey to show for my efforts. In a perfect world, I figured my hive would produce 30 or 40 jars of honey every year to give away to friends and an additional 30 jars to sell for $15 each to cover my expenses. Those were my two simple objectives. But if I eliminated the first year of beginner's-luck honey and looked at my honey yield from when I took over maintaining my own hive, well . . . let's just say I couldn't exactly call the back deck of my float home a profit centre. It was more like a sticky money pit.

Let's take a closer look at the first objective: gifting. In terms of giving jars away to family, friends, and colleagues, there is no better gift. Most of us occasionally get invited over to a friend's place for dinner, and it seems customary to bring a bottle of wine or flowers. A nice one-pound jar of honey wins out over a bottle of vino or lilies every time. Honey is extremely popular. I have never experienced anything but sincere gratitude and praise after handing over a jar of home-grown golden syrup recently harvested from a fresh comb.

I have plenty of experience trying to create housewarming offerings. My homemade wine tasted awful, and the homemade beer I once brewed had a nine-inch foamy head, making it impossible to drink. The lettuce and radishes I grew from seeds and threw into a salad I once brought to a potluck party were puny, deformed, and full of small wormholes. Mushrooms from the forest make iffy appetizers because they could well be poisonous. With gifts like these, it's almost guaranteed your hosts will not invite you back. But there is no such thing as a jar of poor-tasting honey. The bees don't screw up; their honey is always sweet and delicious. It exceeds all expectations in the wholesomeness department, and it never goes bad. People perceive honey as having high value—jars of the size I gave out to friends from the first year's yield sell in stores for over $15.

When I took over the hive, I thought, unrealistically, that I'd sell my excess honey. My imagination wandered off to a cute little roadside honey stand, kind of like the Kool-Aid stand I had as a kid. But alas, I have not sold an ounce of honey yet, and, come to think of it, that roadside Kool-Aid stand never made any money either.

Despite the start-up expenses and elusive profits, beekeeping is still a worthwhile hobby and a great thing for the environment. However, feeding birds is good for the environment too, and a bag of birdseed sells for about $12. I'll cut to the chase: getting started as a beekeeper costs about $1,000. And that's for only one hive. Get two hives, though, because then you don't have to go to your sister's place to steal her frames of bees.

Here's what you need to buy to get started:

🐝 Wooden Boxes ($150)

This is the mini high-rise condo bees live in. You will need three to five boxes. Bee supply stores sell them unassembled and building them is easy, but *easy* is a relative term. Easy for whom? I barely passed Mr. Chambers' grade-eight woodworking class and was one of the few students not allowed near the electric band saw. I failed an assignment in class to build a footstool. It's a good thing that beehive boxes are easier to build than footstools. They have interlocking wooden dovetail edges that fit nicely together. All you have to do is line up the edges and then bang them together, using small nails and glue to keep them in place. This involves a level of skill only slightly beyond putting together Lego blocks. After I build my bee boxes, I like to paint them wild colours using mistinted paint from the thrift store. Traditional beekeeping white, which is often used because it reflects light, is oh so boring. My boxes are a mix of peach, orange, and turquoise and have goofy cartoon bees painted all over them because my girls are special, and I want them to feel special. Their high-rise condo is the grooviest in Western Canada.

Frames to Go in the Boxes ($200)

These are the separate walls in the bee condo. Each box stores about 10 frames, so you need about 50 to get started. In case you are doing the math, the frames cost about $4 each as they are made out of cheap wood and stuck together with small nails. The frame foundations are cool to look at closely because each one has thousands of slightly raised, six-sided plastic cells. These cell imprints get the bees started in building the wax comb required to store honey and raise baby bees. They are the blueprints for the hive. In nature, hives grow in tree branches and hollowed-out logs, and so they grow into weird, wild shapes and configurations. Beekeepers want to regiment, control, and standardize the hive. The best way for us to grow the hive to our specifications is to give the bees a floor plan and to confine the hive's growth inside four box walls. Don't go looking for frames and boxes on Craigslist. It is not a good idea to buy used bee equipment, as it may be infested with microscopic spores and diseases.

Beekeeper Suit ($125)

They come in white or white. When I got my new suit, I took out some fabric paint and wrote my name and, because I was so smug and full of myself after that first year's huge haul of honey, a dumb slogan in big, bold letters on the back. I advertised that I was "The Man, the Myth—the Beekeeping Legend."

Everyone at the outyard must have thought I was an idiot. What kind of guy shows up in a pristine bee suit and a slogan like that? I have had a lot of laughs over it, and it's nice to at least have your name on your suit, since they all look the same. In the next chapter I'll expand on why it is important to have a good beekeeping suit. I learned that I probably should have spent more time selecting mine than drawing on the back of it.

Beekeeping Gloves ($25)

Thick leather covering your sensitive skin from fingertips to wrists will give you the confidence you need to reach into your hives. These rugged gloves could easily handle a rope tow and could double for ski gloves if you were really in a pinch.

Smoker ($50)

You may have heard the expression "smoking the hive." Basically you need a way to introduce smoke into the hive in order to keep the bees from getting agitated and ready to sting you when you do your inspections. You could maybe just stand there holding a newspaper on fire, or you could adopt the bad habit of smoking cigarettes and puff away while going through your bee frames. But both of those are bad ideas for hopefully obvious reasons. A beehive smoker is a metal container; it looks like the oil can Dorothy carried around in her basket for the Tin

Man in *The Wizard of Oz*. The can is attached to bellows. You deposit wood shavings, burlap, dried pine cones, and whatever burnable items you have on hand into the container, then light the ingredients so they smoulder and create smoke.

There are two explanations for how smoke affects bees. The first explanation is scientific. In a nutshell, the chemical composition of smoke masks the alarm pheromones released by the guard bees or bees that are unfortunately injured during a hive's inspection. Sadly, it is inevitable that sometimes bees get injured in the beekeeping process. Opening the hive's lid, moving frames around, and sticking your thin metal beekeeping tool into the hive results in bee injuries and, in some cases, death. I try to avoid this, but when it does happen I resignedly call it collateral damage.

I like the second explanation about the smoke's effect better: the bees smell the smoke, deduce there is a forest fire, and fear it will soon destroy the hive and its store of honey. The smoke stimulates a feeding urge, and the bees all go to the honey pantry to eat like crazy. After they gorge themselves, their bellies become bloated, making it difficult to make the necessary flexes in order to sting. Both theories are considered legit in the beekeeping community. Either way, it is a good idea to have some smoke on hand when entering the hive. Plus, dispensing the smoke is mesmerizing and Zen-like and makes inspecting the hive more ritualistic and mystic. Maybe I'll put some incense in next time.

🐝 A Couple of Good Books on Beekeeping ($50)

This is not one of them, but there are heaps of them out there. I recommend two: *The Backyard Beekeeper: An Absolute Beginner's Guide to Keeping Bees in Your Yard and Garden*, written by a bee zealot named Kim Flottum, and *The Beekeeper's Handbook*, by Diana Sammataro and Alphonse Avitabile. If you ask three different beekeepers the same question about your hive, you may just get three different answers; however, between these two books, you will pick up most of the fundamentals.

🐝 Bees ($200)

If you are going to keep bees, you need to get bees. Here's where it gets weird. Remember that Hawaiian queen that immigrated 2,700 miles to my hive? I have never been able to figure that one out. What's wrong with homegrown queens? Why did mine have to come all the way from Hawaii on an Air Canada flight?

When you go to the bee supply store to buy an entire hive of rank-and-file bees, they come in a long, round cardboard tube called a nuc (short for nucleus colony). It contains the queen and thousands of workers. After buying one nuc, you unplug one end of the tube and pour the colony of bees into your empty hive boxes. All the bees, including the queen, spill out into the boxes and frames. And so they move in. Well, it's not quite that simple. The nuc, as the name would imply, is the heart of the colony and will grow and grow as the queen throws her reproductive organs into high gear and the nurse bees do their duties. The most common place for the nucs to come from, though, is—get this—New Zealand! Up until now, everything I have learned about bees relates to how sensitive their systems are to light, air pollution, hive positioning, radio waves, temperatures, and any subtle change in the environment. I just don't understand how you can ship these poor little creatures halfway around the world and expect them to adapt. But they seem to. Hey, my Hawaiian queen was working out, so who was I to knock imported bugs?

Jars and Labels ($100)

Two quick pieces of advice: Buy small jars and give honey away sparingly. You never know how long it will take until you get your next haul. The second tip is to get creative with the label. Come up with a cute name and design for your honey operation to enhance its appeal.

Miscellaneous ($150 to infinity)

There really is no limit to the amount of hard-earned dough you can lay down for the holy purpose of keeping your bees. A beekeeper cannot live without the infamous, generically named beekeeping tool. It's a little gadget we carry, a bit bigger than a can opener, to help us move frames, cut wax, and separate boxes. It is as easy to misplace as my reading glasses. If you throw in an uncapping fork tool, a synthetic beehive brush, and a mini beekeeping entrance feeder, you can obtain a little package online for about $35 that resembles the assortment of utensils usually found in a kitchen drawer. The sugar to feed the bees and the chemicals to ward off diseases and to kill mites will cost you another $100. Once you get into beekeeping, as with any hobby, you'll discover there is no shortage of accessories and handy gadgets you can buy, such as tiny doormats for the hive's entrance to brush pollen off their wee legs. I am serious. You can also spend your money on a mouse guard, a small metal gate that fits over the hive's entrance to keep pesky rodents at bay. I wish I had one for the inside of my float home, too!

Adding up the total cost of materials and figuring out my beekeeping return on investment has certainly been an eye-opener—indeed a bit shocking. During the first year I raised bees on my own without as much assistance from my expert sister, brother-in-law, and girlfriend—to whom I give most of the credit for the original 100-pound honey haul—I personally extracted about five pounds of honey. That is the equivalent of five jars. Each jar, then, cost me about $200.

I am left wondering if friends who invite me over for dinner might prefer me to hand them 20 crisp $10 bills instead of a jar of honey as a host/hostess gift.

Show me the honey!

I'VE LOOKED AT BEES FROM BOTH SIDES NOW

Beekeeping is all about function and not about fashion. It is impossible to look sexy, attractive, or important while wearing a bee suit. Though you will get a strenuous workout while beekeeping and it is great for your cardio, don't expect to look even remotely sporty or fit while wearing the suit. In full regalia, beekeepers look like nerds. While wearing my extra-large suit, I feel like a combination of astronaut, clown, and Pillsbury Doughboy. But, safety first. By wearing a quality suit with a sturdy, well-made metal zipper that properly fastens on the head veil, thick leather beekeeper's gloves pulled up high to your elbows, and inset fabric elastics properly sealed around your ankles and wrists, you are well on your way to keeping the bugs where they belong: outside of your suit.

Even so, four of the suit's possible entry ports are the two leg bottoms and the ends of the two sleeves. Most suits have Velcro fasteners and elastic cuffs on the wrists and ankles expressly designed to block out the bees. If you are not satisfied with the design of those protective features, then good old-fashioned thick grey duct tape works extremely well to wrap around your limbs as a double safety precaution. Duct tape, I have learned, is an essential part of beekeeping. Duct tape is an essential part of modern life; it has a light side and a dark side, and it is the force that holds the universe together.

There is a fifth port of entry, and that is the neck zipper. The first zipper was showcased at the world's fair in 1893; however, it didn't take off until an enterprising front-runner in fasteners named Gideon Sundback made some modifications to the prototype in 1913. Since then, zippers have helped to secure, fasten, and close a variety of modern conveniences, from luggage to sportswear, but most of the time we never give zippers much thought. An errant zipper rarely creates panic, pain, or paranoia. Zipper flaws or accidents in daily life at worst result in a partially opened jacket on a cold day. A partially open zipper fly may cause an occasion to blush. But nowhere is a zipper malfunction more dangerous and alarming than in the world of beekeeping.

As far as I'm concerned, the zipper is *the* most critical detail when selecting a beekeeper's suit. A suit with a strong metal zipper to attach the all-important sealed head veil is paramount. Since a large part of this chapter will deal with the efficacy of a 12-inch zipper (talk about a boring subject), it is really important you understand exactly where this zipper is located. Imagine the round collar on the top of a T-shirt; then imagine that collar as a zipper, and that it attaches to another zipper fastened to a paper bag–shaped hat the size of your head, only the paper bag is really a protective veil with screened mesh on the front of it, similar to the mesh on a screen door. The veil with screened mesh is the only thing between your face—eyes, nose, lips, rosy cheeks—and a tribe of potentially angry, unsettled bees. Get the picture?

To fortify this getup, one must also wear the proper clothing underneath the suit: thick jeans and a hoodie. And don't forget your feet! Never do what I once did accidentally and wear a pair of open-toed sandals while tending a hive. A big juicy stinger planted firmly into my big toe was the best reminder not to do that again. Woollen socks and ankle boots are the mandatory finishing touches. Wearing a baseball cap under the veil, which makes you look even dorkier, will prevent facial stings. But even with all of these unfashionable clothing precautions, are you actually completely protected from getting stung?

The answer is no. I have had several wardrobe malfunctions in this department and, as such, speak as an expert. Bees have made it through my suit on four occasions and even snuck inside the screened head veil. Each horrifying trespass can be directly attributed to a word that begins with the evil letter z. When it comes to bees, bears, and people, there is no such thing as an impenetrable outfit or enclosure. There are always small holes and tiny openings to be discovered. I have seen that determined wasps, ants, and mice will eventually gain access to the best-guarded, most-sealed beehive. And I am sure that one of these days we will go up to the outyard and a bear will have broken through the electric fence. As for people, there are countless examples of bank heists, museum break-ins, and forced-entry home burglaries that are a testimony to ingenious ways of getting into places we are not supposed to be.

Since bees have the capacity for reason and the ability to communicate with one another, I believe they figured out long ago how to outsmart modern protective-clothing manufacturers. Scientists believe bees have been around for millions of years. In a paleontology book, I once saw a picture of a bee found embedded in a translucent piece of amber rock from a mine in Burma. The carbon dating of the amber nugget was 100 million years. Something tells me that any bug that has been on the planet for that long has had to figure out more difficult challenges than penetrating a cheap zipper. Compared with surviving the ice age, sneaking into a bee suit is a piece of cake.

One place that will test your suit's mettle is the outyard. In the outyard there are hundreds of thousands of bees coming at you from every angle, conspiring with one another to figure out a way to get into your bee suit. In each instance when a bee has gained unlawful access, I swear I did the zipper up properly and even had Jeannie double-check it. But after 20 or 30 minutes of vigorous beekeeping movement, the zipper somehow came partially undone in the middle. The problem, of course, was that the suit had a cheap plastic zipper. When the zipper of a bee suit splits open at the back of your neck, it creates a tiny opening that acts as a welcome sign for bees. Since that malfunctioning zipper hole was out of sight, I was unaware of the danger, which would eventually lead me to a catastrophic and painful meltdown—every beekeeper's worst nightmare. It's hard to believe the pain and anguish I experienced was due to three or four separated zipper teeth.

Few things are more terrifying than having a bee trapped on the inside of the protective veil. You have to remember that once a bee is inside that cramped canvas cavern and crawling on your face, there is zero chance to get it out. You can't simply unzip the veil and remove the bee because in the outyard there are too many other bees zooming around. So you just have to whack the invader with your hand, bashing the side of your face as you miss the bee. It keeps buzzing around, getting more and more agitated while trapped within an inch of your sensitive skin. The buzzing noise blasts in your ear the whole time, amplified tenfold by the acoustics in the hooded veil.

That seminal moment of realizing that you have "a bee in your bonnet" is terrifying and shocking. Consider this: while at the outyard, because the air is thick with hundreds of thousands of bees flying around, you will always have between one and two dozen bees crawling on the outside of the see-through mesh veil. They are one inch from your eyeballs, nostrils, and lips. Their six legs are facing you, and their tiny stomachs are rubbing against the screen. But you don't have to worry because they are safely on the other side of the protective mesh. They are always there, but not a threat. Then, imagine the moment when you spot the reverse image of a bee—you spot the *back* of one. It is crawling on the same mesh with the rest of them, but its *wings* are facing you. You realize it's on . . . the inside.

One day Jeannie and I were at the outyard and a nice lady named Judy, who was up there tending her own hives, suddenly stopped, looked hard at me, and said, "Dave, don't move. There is something wrong. Don't panic, but I think you have a bee flying around on the inside of your headgear."

She might as well have stuck a revolver to my forehead. I launched into hyperanxiety and lost my breath as she swiftly and deftly grabbed the loose head veil cloth and squashed the bee inside the mesh to death with her gloved fingers. I narrowly escaped the worst of all stings: the facial sting.

The second time a bee snuck past the zipper and got too intimate with my face, I panicked and ran like hell while hitting my head with my hands over and over again. I must have landed a lucky blow because the bee died before it could sting me.

The next time I visited the outyard, I carefully put on the suit's head veil and tightly zipped it up. Then, as an extra precaution, I duct-taped the zipper shut. Even with this extra safety measure, a bee penetrated my defences and got inside. I didn't see this bee; I heard it, because it flew directly into my ear. It actually crawled inside my ear. I panicked. Again, I ran like hell, violently slapping myself silly in the ear with both hands and arms. I hit myself so hard I almost popped my eardrum. Thank goodness I got this bee before it got me. I smacked my ear 20 times in a row with so much force that I somehow squished it. After we wrapped up that hellish outyard hive inspection, I asked Jeannie to check the back of my suit. The cheap plastic zipper had come undone, again. The bee, apparently intent on becoming an ear dweller, had somehow manoeuvred through the duct tape, then through the opening in the zipper. What the heck she wanted to do in my ear I'll never know. Maybe she mistook my ear for a small six-sided hive cell and sensed that it already had some wax in it.

After that string of accidents, it was time to take charge. I complained to the bee store where I bought the shoddy suit, and they exchanged it for a new suit with a hardy metal zipper. I could now venture back to the outyard with confidence, knowing my days of facial assaults were behind me. I test-tugged the new zipper on the suit and could not tear it apart. It was bulletproof!

As it turned out, the fourth attack on my face at the outyard, which happened while wearing that new suit, was the scariest and most miserable of them all. It was one of the hottest days of the summer. To this day, I remember it as the day I almost gave up on beekeeping entirely.

When I am in the bee suit, I wear jeans and a heavy sweater underneath. Being sealed up in the suit on a hot day wearing all that clothing and lifting heavy boxes can feel like being inside an oven. The pea-sized beads of sweat that flow from your brow and down your forehead probably weigh about the same as a honeybee. Come to think of it, the little drops are about the same size as bees too. On those "turn your oven up to 350 degrees" days, gravity pulls the beads of sweat off your forehead and down your cheeks. On that fateful day, I could feel warm drops of sweat slowly dribbling down my face at about the same speed that a honeybee crawls. I couldn't tell if it was sweat or a bee, or even several bees crawling on my chin, lips, and nose. Three times I have had bees crawling on my face and it always felt just like this. If Alfred Hitchcock were still alive today, I would send him this book's manuscript and suggest it as a follow-up to his classic movie *The Birds*. My horror flick, however, would be called *The Bees* and would open with a scene of me in the outyard sweating it out in a paranoid state of incoherent, trembling fright. The camera would pan from the sweltering hot hives to me, pale-faced behind the mesh, and then dissolve into a close-up of a nervous twitch on my glistening cheek just before I scream in horror as a giant bee slowly inserts her massive razor-sharp stinger into my soft flesh. The scene would be in black and white and slow motion, with an eerie solo violin soundtrack.

That calamitous afternoon at the outyard was a real-life thriller. The bees were aggressive and mad because it had rained earlier and was so hot. I had the suit on for less than five minutes when I got stung on the wrist. Bull's eye! Some little suicide maiden managed to insert her pointy-ass stinger through the cotton canvas fabric just below my left hand. Distracted by the pain, I walked away from the hives for a moment to catch my breath. A cloud of angry bees followed me. The pheromones released after the dying bee had planted her first stinger in me alerted thousands of other bees to come after me and attack. Pheromones in this case are like an SOS distress call. They signal: "Danger is present, girls. ATTACK the big white unfashionable geek predator." My suit was soon covered with hundreds of berserk bees stinging the suit fabric. After a minute of being a human dartboard, I retreated to Jeannie, and she couldn't believe how many stingers were stuck in the suit. We estimated over 100! Without the suit, I would have been dead.

I would not call it quits, though, never, not me. A few minutes later, I was back to work, assisting Jeannie in tending the hive. That's when it happened. I got a solid direct hit square in the face. It wasn't from the inside though; the new zipper was working. It came from one of the dozens of bees crawling outside of the suit on the see-through mesh. It is important to keep that mesh veil from touching your face—hence the baseball cap. The brim of the cap holds the mesh away from your face. I had an old Seattle Mariners cap, but it was back in my car at home. I know, I should have to wear a dunce cap as punishment. But another form of punishment was swiftly administered. With the mesh touching my face, a bee planted her little bee bum, armed with an atomic-powered toxic stinger, right through the screen and into my chin. The venom exploded under my skin, and the side of my face swelled up immediately. I looked like I was hiding a tennis ball inside my right cheek.

The vicious circle continued with more sos pheromones released from the latest sting, attracting thousands of new attacking bees, resulting in hundreds more stingers planted into my suit. The bees were ganging up on me. Each stinger sounded the alarm to come and kill me. I felt like such a loser. Why does this always happen to me? Jeannie and my sister get stung only occasionally. I was now a human pincushion.

With two stings in less than five minutes, I needed to take some time out. I left the fenced outyard and went to sit on the ground next to Jeannie's truck. I sat there in the dirt with a thousand of my closest new "frenemies" swarming around me in the hot, humid air. I wondered if I had kept the receipt for the newly exchanged suit. Then that awful nervous paranoia entered my consciousness again. Was the slow crawling sensation on my face a bead of sweat or a bee? I slapped myself a few times and the creepy feeling stopped. I closed my eyes, took a deep breath, and tried to relax. But the sensation returned, this time on my lower lip, then my upper lip.

Then I had an unsettling thought: sweat doesn't drip upward. I felt six tiny legs crawling on the edge of my nostril and slowly up the bridge of my nose toward my left eye. I shot up off the ground and screamed for Jeannie, and then began running as fast as I could away from the truck and the outyard and the concentration of bees. As I ran, I felt the bee on my temple, then it touched down on my eyelid, and then it settled on my forehead. I relentlessly bashed myself in the face with my hands. When I was so exhausted I could not run anymore, I fell onto my knees to catch my breath. There were still too many bees around me in the air so I couldn't take off the headgear, but no matter how hard I punched myself, I couldn't kill the bee inside the veil. It just kept crawling all over my face. Then the sensation of the bee stopped. There was no more buzzing. Had I killed her? Had I imagined the whole thing? Was my brain playing tricks on me? Because I had become so paranoid, because of all my bad experiences up there with faulty zippers, was I starting to lose my mind? I felt like I was about to throw up, and what could be worse than barfing in a bee suit?

Anyone arriving at the outyard during the moment I bolted in terror would have seen a pretty funny sight. Fans of Charles Schulz's comic strip *Peanuts* will remember the character Pig-Pen. Pig-Pen always had a cloud of dirt surrounding him wherever he went. I had my cloud of bees. They must have been bees of the Usain Bolt variety, because no matter how fast I ran, they kept up with me.

Just as Jeannie arrived after running down the 100 yards from the hives, I looked up to the corner of the head veil and saw two big bee wings facing me. The possibility of another sting on my face scared the living hell out of me. I was still in a bit of shock after the first two. Luckily, Jeannie had her wits about her and quickly reached out her hand. She grabbed the veil, crumpled the fabric where the bee was, and killed it. I got up off my knees and ran farther down the bumpy dirt forestry road. When I sensed that instead of hundreds of bees surrounding my head, there were only dozens, I frantically unzipped the claustrophobic, overheated veil and pulled it off my head. I tried to breathe in as much fresh mountain air as I could to calm myself down. Crestfallen, I sat in the dirt, catching my breath and hoping to slow my accelerated heart rate. I took off my left glove and touched my bare hand to my cheek and lips. I could feel the throbbing and the swelling. Beekeeper down!

Because of my panicked quick release of the head veil, I will never really know how that last bee got in. I tore the headgear off too quickly without checking if the zipper was properly fastened. Did she enter through a hole in the zipper or a leg or arm opening? Who cares—she was dead now, and I had more important things to worry about, like the venom from the sting on my chin spreading to my neck. Not to mention the two dozen bees still circling me, refusing to go back to the hive.

That's when we decided to call it a day. Jeannie finished loading some frames into the back of the truck by herself and then climbed into the front cab still in her full beekeeping gear. As I opened the passenger door to join her she said, "You can't come in here; you still have too many bees swarming around you!" She was right. I just couldn't shake them. So I went to the back of the pickup and crawled into it with all the empty boxes, sticky sweet-smelling frames, and the various pieces of beekeeping equipment. Sitting on the hard metal floor of the truck bed, I hardly felt the jarring bumps because my mind was more focused on my now grapefruit-sized chin. We couldn't get away from that torturous outyard and back to a bee-free zone fast enough. I still had a dozen hangers-on even as Jeannie slowly accelerated. It wasn't until the grade of the road improved and Jeannie was able to accelerate to about 15 miles per hour that we were able to lose the last stubborn bees circling around me. I couldn't have been happier to leave them in our dust. A few more miles down the road, Jeannie let me back into the truck's front compartment, and after 40 or 50 more miles, we were back at Jeannie's place, where I could finally relax.

I took the suit off and carefully inspected the zipper. I zipped and unzipped it three times, and each time the teeth fell into place with military precision. How the hell did that bee get in?

I held the suit above my head to inspect the fabric with the sun shining behind it. Maybe it had a tiny hole somewhere? As I lifted the suit, the bee Jeannie had killed fell out of the veil and onto the ground. The lifeless body landed on its back, legs reaching for the sky above. Rigor mortis had set in to those six motionless legs, and her stinger was still perfectly intact. I found it ironic that after such a valiant and heroic penetration of my bee suit, she never had the chance to unleash a lethal stinger blow to my face. She had come so close.

With the bee lying there on her back, I observed the symmetry of her body parts and the beauty of the yellow-and-brown colouring on her stomach. Then I carefully turned her over so I was looking down onto her wings. I gasped, the hair on the back of my neck stood on end, and my knees felt weak. Just the sight of her wings facing me put me into shock, a particular kind of PTSD: pollinator traumatic stress disorder. It occurs when you know the excruciating pain and humiliation that awaits when you spot the terrifying sight of a set of wings facing you and know there's a bee crawling on the inside of your veil. It's basically a living hell. Having observed bees from both sides now, I prefer to see their bellies crawling outside my mesh veil.

WASPS, MASHED POTATOES, AND AN ETCH A SKETCH

One September day, I got all kitted out in my bee suit, grabbed my hive tool and a roll of duct tape, and went to inspect my hive. A sad sight greeted me: 100 dead bees at the entrance. I opened the hive only to find a baseball-sized pile of bee corpses. Worse yet, I pulled out a frame to see dozens of wasps stealing my honey.

Remember when I replaced the queen in my hive not once but twice, and after a few false starts she finally began laying eggs? I got all excited over fresh brood and hope sprung eternal, but in the world of beekeeping a lot can happen in a few months. In the end, the second new queen turned out to be "all crown and no cattle," earning a D minus in the reproduction department. I have no idea what the new Hawaiian queen was doing in there, but she did not lay nearly enough new eggs to kickstart and sustain my hive. A couple of months after the new queen was installed on her throne, every inspection revealed fewer and fewer bees, less brood, reduced stores of honey, and increased mites and disease. Because of the queen's infertile nature, my faithful hive—which had produced the award-winning Houseboat Honey—was going steadily downhill. The wasps were taking over.

Wasps are so darned intuitive and smart they can actually detect the scent of a sick hive. It is impossible for us humans to smell it, but a sick hive exudes olfactory evidence of weakness, illness, and stress. When wasps detect this telltale scent, they attempt to enter the hive. Once they overpower the guard bees, they mark the outside of the hive with a pheromone, then leave to bring back hundreds of reinforcements. If the hive radiates a healthy and robust smell, the wasps only work the base around the hive, scavenging and eating up the occasional dead bee lying there. These are the kinds of guests you don't want to come a-calling at your hive.

When wasps do attack, they do so with a vengeance. My hive of sick and diseased bees was unequipped to stop them. The once reliable and robust guard bees were now listless, battle-worn invalids, unable to thwart the ruthless invading wasps. It was no match—I actually watched the wasps fly up to the hive's entrance, bite the guard bees to death, and march in. Injured bees, dead bees, and bee parts littered the entrance. You can tell once a bee has been in a bad fight because it loses its fuzzy outer coat. It usually also loses a leg or two.

Even reducing the size of the hive's entrance was no longer helping. The wasps bullied their way in. After gaining entry into the hive, they came back on a regular basis to steal more and more honey and larvae (baby bees), which they take back to their nests and feed to their wasp larvae. Every day, hundreds of wasps were dispatched from their nests to overthrow my hive. Every day, these unruly heathens were eating more and more of my bees.

While all this was going on, my useless Kona Queen was probably somewhere in the hive hula dancing to Don Ho records. Since she never laid enough eggs, we didn't have enough healthy young bees to take up the charge and continue the vibrant circle of life in the hive. Between the low birth rate from the deadbeat queen and the high slaughter rate by the wasps, my two bee boxes were housing about one-quarter of the bees they had when the hive was healthy. The remaining bees, sadly, didn't have the same pep and zip as they had before. What was really troubling was that winter, a time when the hive needed to be at maximum strength and population to survive, was just around the corner. I had one last chance to save the hive. I needed to stop the wasps.

It was time for me to bring out the heavy artillery. Come hell or high water, I wasn't going to let those damn marauding killer wasps rob and kill my poor little bees. The weary bees and I had worked too hard for what little honey was left in my decrepit hive to simply have it ripped off from right underneath our collective proboscises. Under no circumstances would the unruly, destructive gang of wasps circling my float home be allowed to pillage the ragtag inhabitants of my weakened, defenceless hive. Not on my watch. Nope. The proverbial buck would stop here. To protect my girls, I sat on my float home's back deck in a rickety black wicker chair next to my beehive, armed with three weapons of mass destruction (WMDs).

The first WMD I used was a plastic tennis racket-shaped electric fly swatter, its killing efficiency increased by the brand new AA batteries I had installed earlier that morning. When the plastic racket's wires made the slightest contact with a wasp, the would-be intruder was immediately zapped to death. Whoever invented this gadget will have to answer to 100 trillion fried bugs at heaven's gate all asking the same question: "What happened?" As the death racket's operator and executer, you need very little skill or experience. It's easy. All you need to do to make contact with an incoming wasp is to wave the racket back and forth in the air as if it were a magic wand. Even the slightest grazing with the racket's highly charged, interwoven silver wires instantly turns a healthy wasp into a crispy critter. The racket's generous "death surface" gives you a good chance of success, and you can sometimes even hear them pop and explode after a direct hit. It's not something I enjoy doing (that much, anyway), but no one ever said beekeeping was all pretty flowers and rainbows.

My commonsensical brother-in-law Len taught me about the next wasp extermination tool: a simple everyday vacuum cleaner. The tip of a vacuum hose has a remarkable suction radius. As with the yellow plastic racket swatter, you just need to get the vacuum's tornado-strength force in the general vicinity of a wasp, and it's back to Kansas for that bug.

Sitting there in my bee suit on the back deck, I had two modern labour-saving pest control devices to protect the hive from brutal invasion. In my left hand, I held the electric swatter; in my right hand, the nozzle of a vacuum, plugged in and running at full power. Isn't electricity a wonderful thing? I preferred the wasp-sucking vacuum method. It was cleaner—no dead bugs dropping at my feet, no snap, crackle, and pop, and no disagreeable odour. But in the back of my mind I wondered what happened to the wasps after they got sucked into the one-way trap and were carried by the powerful suction motor down into the vacuum cleaner's storage bag. Could they live down there? Could they *breed* down there? The vacuum bag would be full of dust and dirt and random detritus, possibly including some tiny bits of food. Ironically, I might have been transferring them into a wasp Shangri-La. After I stowed the vacuum in the closet, the concealed bag would be dark and quiet and full of food crumbs. Just the environment wasps love. Could the 30 or 40 wasps I sucked in that morning form a colony deep in the bowels of my Hoover, multiply tenfold, and escape into my home in the middle of the night? That, dear reader, is the stuff of nightmares.

So, following my killing spree, I pulled out the ever-handy duct tape and plugged the vacuum's nozzle, their only possible way out. I replaced the vacuum bag the next day. The modern disposable bag had an only two-inch-diameter hole and I couldn't see inside. I wasn't about to grab a flashlight and stick my eyeball up to the hole, allowing a wasp the ultimate revenge of tearing off a chunk of my cornea. Who knows what happened to those wasps that were sucked in earlier? Something tells me it wasn't a happy ending; thus, dead or alive, they got taken out with the trash.

The final wasp destruction method I used that morning was a simple, old-fashioned pair of thick beekeeping gloves. The leather that covered my hands and fingers made it impossible for wasps to bite through while I squished them.

I needed to quickly decide in any given moment which of the three execution methods to use, depending on whether the wasps were flying or crawling. Mid-air flights were best intercepted with the electric death racket. Alternatively, rowdy wasps that landed on the hive's front entrance looking to break in past the guard bees were immediately dispatched to Hoover heaven. If a wasp was walking around on the flat metal surface of the hive's lid, sometimes it was easiest to reach my hand out and squish it. One conclusive scientific observation I made after my first hour-long extermination study was that wasps react slower to a heavy gloved hand thunderously descending upon them than bees do. Bees have two big eyes—plus three small ones—that wrap all the way around their heads with about 6,000 lenses to spot a swift impending blow from any direction. Maybe wasp eyes have only 4,000 lenses and therefore less peripheral vision. I don't know; I'm a beekeeper, not an entomologist–ophthalmologist. What I do know is that in the hour I sat there, I hoovered 17 wasps, zapped 11, and squished 21 for a total death count of 49.

At the peak of summer, the average wasp nest holds about 5,000 potential marauders. Did you catch that? Wasps live in nests and bees live in hives. If you have a tough time recalling which is which, just do what I do and remember that "nest" rhymes with "pest." Regardless of what you call a wasp's home, it's easy to do the math and figure out how many wasps I killed that morning relative to the wasp nest's population: less than 1 percent. I figured I would have to sit there for at least 100 hours or about four straight days to neutralize the threat to my girls.

Unlike bees, wasps don't winter. Only the hardy queen wasp lives to see the next calendar year. The entire wasp colony dies during late fall, and the queen flees. She finds a tiny crack in a rock or tree to hide in and hibernates for five or six months. She emerges in the spring to build an entirely new nest from scratch. I thought if I could keep my beehive alive for another few months, the wasp problem would simply go away.

Sitting out there on my deck in my protective suit, I would occasionally kill a bee in the process of trying to kill a wasp. I'd shrug this off to the uneven suction radius of the vacuum cleaner or the bulky, oversized voltage screen of the electric racket. Once or twice, my heavy-handed leather glove accidently landed on the wrong insect species. You may think that inadvertently killing my own bees would bring on a tremendous amount of guilt, conflict, and remorse. Not really. Here's the way I look at it: once a wasp makes it past the guard bees into the hive, it becomes a maniacal killing machine. I observed many a tussle between my bees and the attacking wasps. The wasps always won. My poor girls are smaller than the wasps and have only one line of defence: to sting once. Wasps have a huge upper hand in that they can sting and bite repeatedly. Like bees, wasps are equipped with a stinger that contains poisonous venom. While a bee can sting only once because its stinger becomes stuck in its victim, a wasp's permanent stinger remains intact and gets reloaded with venom to sting again and again. Wasps also have sharp teeth or mandibles to hold and bite off pieces of prey, which they carry back to their nests. I read somewhere that after a single wasp invades a hive, it can easily kill between 20 and 30 bees. As the pragmatic and conscientious steward of my hive, if I accidently sucked an errant bee into the vacuum, it was for the greater good. Each wasp that got fried, squished, or dragged into the dust bag resulted in 20 or 30 bees getting to live for a few more weeks. Some bees were sacrificed, but in the end I did what I had to do. I had their backs. Not that I enjoyed this.

I did make a real effort to distinguish between bees and wasps in order to kill only the bad guys. That particular morning, I had to be super observant and totally focused. Wasps are slightly bigger than bees, sleeker, and a brighter yellow. They are sometimes called yellow jackets for that reason, and their contrasting black stripes seem to stick out more than the muted two-toned friendly bees. But the telltale sign that distinguishes wasps from bees is the way they fly, and since my electric-racket strikes were aimed at airborne wasps, immediately identifying them in flight was important. To increase my identification accuracy, I needed to have my 2.5-magnification reading glasses on under my bee suit's veil. Hundreds of wasps and bees flying frantically around the immediate vicinity of the hive are all jumbled together and to the untrained eye look like a cloud of bugs. A hobby executioner like me needed a way to quickly distinguish friend from foe. However, it was hard to clearly see their size, shape, and colour through the bee suit's mesh veil and my always-dirty reading glasses. That's where the importance of flight patterns came in.

If you carefully study the flight patterns of both insects as they approach the hive, you will see that bees tend to arrive directly in a more orderly, circular fashion, and they are less jittery than the wasps. Wasps have a completely different way of buzzing around. They scan back and forth when they fly. If you were given an air traffic control computer readout of their flight patterns, it would look like the dark grey right-angle lines a child creates with an Etch A Sketch. Wasps are also more aggressive and quicker in flight than bees, making them a bit harder to pick off with the racket. But they almost always turn at perfect right angles with military precision.

I first became aware of wasps at the age of seven. To know them is to dislike them. In the summertime, we would occasionally have family dinners outdoors on our veranda, especially when my granny came to visit from New York. No sooner would my sister, my mom, my granny, and I be seated around a wobbly old wooden table, with the sun setting in the background, ready to dig into a delicious meal of roasted chicken and mashed potatoes, than a lone wasp would show up out of nowhere and ruin our lovely dinner. At the first sight of a wasp, Miriam would scream, Granny would almost faint, Mum would run inside to get a rolled-up newspaper to swat at it, and I can't remember what I would do. I probably just sat there daydreaming with my finger up my nose, wearing my cute red Canadian Mountie outfit from Granny, quietly observing and thinking that maybe in 50 years I would write about the winged intruder.

And it happened, because sure enough, five decades later, there I sat, alone on my back deck, clutching a vacuum nozzle in one hand and a plastic electric tennis racket in the other, wearing a beekeeping suit with my reading glasses on underneath my face net. I had a lot of time to imagine writing about these hazy, trivial childhood details zigzagging through my brain like wasp flight patterns. Remembering those innocent, carefree summer dinners on the veranda, I envisioned the way the wasps flew and ruined all of our meals. My mind drifted back to our old white-and-green house on the corner of 11th and Crown, where a perennial wasp nest hung like a lantern in the cherry tree in the backyard. I vividly recalled the erratic movements of the wasps—the sharp right-angle turns, the frenetic reconnaissance before they landed. I remembered the horrible buzzing sound. Finally, I recalled the most disgusting detail of all: the likelihood that before arriving at our dinner table, the wasp had just landed on a big hunk of our German shepherd's poo. And, covered in Thor's feces, it was now crawling all over our savoury chicken. Is it any wonder I can't stand wasps? As I sat guarding my bees that morning, my resolve to kill as many wasps as I could only deepened. I had a score to settle.

Suddenly I realized I didn't have any more time to daydream. I had an important appointment downtown that morning. As bad as it felt, I had to abandon my weakened and fragile bee colony, forcing them to fend for themselves for a few hours. I drove into town picturing the never-ending onslaught of wasps my bees were facing on their own. To distract myself from these sad thoughts, I turned the news on the radio up full volume. With me not there to fend them off, the wasps would simply overpower the guard bees, probably killing many in the process. They'd march into the hive and continue their killing spree as they bullied their way up through the frames toward our sealed cells of honey and the innocent newborn baby bees. Nature can be so damn cruel. It truly is a dog-eat-dog or, more accurately, a bug-eat-bug world. Wasps were a lesser life form than bees—a more sinister life form. Although the two bugs have a lot in common, bees became vegetarians millions of years ago and, as such, morphed into what I consider to be a gentler, kinder species. There I go anthropomorphizing again, but it's just how I feel.

Let's compare the two insects. Animals need protein. Bees get their protein from the pollen of pretty flowers. Wasps get their protein from showing up uninvited to family barbecues, and from killing and eating my bees. Bees build elegant castles out of wax, moulding them to serve their every need, creating intricate birthing chambers, storage pods, and small living quarters. Wasps gather grass and bark, chew it and mix it with their sticky saliva into a pulpy fibre, and then piece it together into shoddy, substandard paper nests. Wasp nests don't need to be well built like beehives because wasps abandon their nests in winter. Wasp nests are nothing more than poorly designed, ugly temporary housing. Bees gently flutter from flower to flower; it's almost poetic to watch them joyfully gather nectar and pollen from sweet clover, thistle, alfalfa, and dandelions. Wasps erratically zero in on dog doo. Bees are prolific pollinators, helping important flowers and plants grow all over the world. Though there is evidence wasps are also pollinators, and they eat a number of garden pests, after watching them snuff out my bees, I still wish they would disappear off the face of the earth tomorrow. It's also notable that the word *wasp* is painted on warships and planes, while the word *honeybee* is lovingly used in children's books.

I needed to eradicate these wasps. Driving back to the float home later that afternoon, I stopped by the local hardware store and went straight to the pesticide section. I purchased three wasp traps for less than $20. Two were of the "just add water" variety; the third one required a tasty recipe of household scraps to tempt the wasps into a one-way prison.

When I got home, I followed the directions on the two ready-mix traps, filling them with water exactly to the blue indicator line on each bag. When it came to preparing wasp bait, I, uncharacteristically, figured it was best to pay attention to the smallest of details. A watery concoction might not be strong enough to attract the villains, and if the premixed crystal liquid bait smell was too concentrated, it might tip them off to something being up. When I had the mixture just right, I jumped into my bee suit and ventured out to the hive with a death trap in each hand. The executioner was back, this time with help from the Woodstream Pest-Control Company. I carefully hung two of their yellow lunch bag–sized plastic traps equal distances to the right and to the left of the hive.

I stuck around for a few minutes, and sure enough, the wasps were all over the traps in no time like black on an eight ball. A few seconds after I hung up the bags, the wasps' flight paths began to divert in that unmistakable jagged traversing pattern, no longer zeroing in on the hive but heading straight to the plastic traps. Once they landed on a trap, they'd sniff, poke, and wander around for 20 to 30 seconds before following the tempting scent to the motherlode of irresistible liquid down below in the bag. Only problem was, unbeknownst to them, to get there they had to proceed through a clever one-way trap door. Content in knowing that wasps were already dying, I went back inside to select the bait to entice more wasps to their deaths.

My first idea for bait to use in the cylindrical-shaped trap was, of course, mashed potatoes. I knew from my childhood on the summer veranda that mashed potatoes were a tried-and-true wasp attractor. What the heck, maybe I'd even throw in a chicken wing for good measure. However, upon further consideration, I decided it was impractical to whip up a dish of mashed potatoes, especially since I didn't have any potatoes or butter. At no time did the thought of putting dog doo in the wasp trap even cross my mind.

Choosing just the right food bait was an important decision, but I didn't need to rush it. I could relax and concentrate on my options, knowing that while I was ruminating over the perfect insect meal, some of the wasps outside near my beehive were already drowning in two big pools of liquid crystal bait. With the two other traps out there already catching wasps, I had bought my bees a little more time.

I read through some old emails from Len, who had rid his hive of wasps the previous summer. He had used rotten sardines as bait—not a bad idea. The instructions on the trap's packaging called for soda pop along with table scraps. I didn't have any table scraps, so I opened a can of flaked white tuna and scraped it into the trap's cylinder. I don't drink soda pop, so I used red wine. I found an old piece of Stilton cheese in the fridge and added that to the mix, thinking it would go well with the wine. Finally, I threw in a banana peel for good measure. I had just set the table for the wasps' last supper. They loved it. They came over for dinner and never left.

The three traps were a success. They killed hundreds of wasps. All three traps were loaded with wasp carcasses, and when I inspected them more carefully, I was astonished that not one bee fell for the trap. Bees are attracted to nectar and pollen, not banana peels, stinky cheese, and tuna. I told you that bees were superior to wasps.

Yet, even with a few hundred dead wasps in the traps and the 50 or so I'd killed earlier in the day, I was losing the war against the 5,000-strong army. But I had one more trick up my sleeve.

My last-ditch effort was a total hive lockdown, kind of like what you hear about on CNN when a building gets a serious threat. Well, the wasps were definitely a serious threat—the wasps' ability to get past the guards, gain access to the vulnerable hive, and go on a killing rampage conjures up some very disturbing present-day analogies. So one night I locked the whole hive down. In the world of modern beekeeping, locking down the hive so nothing can get in or out can buy some time and may even rid your hive of wasps if you do it properly. As I prepared to lock down my hive, I briefly wondered if CNN would send Anderson Cooper out to cover it.

Here's how it works: At night all the bees return to the hive, and, theoretically, the wasps go back to their nest. Like us, bees and wasps have circadian rhythms; thus, they sleep in their own homes. Late at night, before you go to bed, you completely seal the hive by blocking the two hive entrances with pieces of foam. I cut pieces out of those colourful swimming pool noodles.

In the dark, with all of my girls safely at home, I shoved a big piece of blue foam into the entrance door and sealed it with duct tape. Now there was no way the wasps could get in the next morning. But since the bees can't go out and forage, how do they survive? That's where a giant jar of sugar water comes in. You leave the bees with a survival kit in the form of a jar of sugar water to feed them and a jar of plain water to hydrate them. Then you leave them locked down for three days and hope the wasps dropping over during that time lose interest in the hive after they discover they can't get in. Hopefully they find some other weak hive and go and raid it instead. You also hope your bees survive while trapped inside.

Well, it sounds good on paper, but here is where the wheels fell off of the plan. It was my own fault. I gave the girls their survival rations and blocked the two entrances. When I was about to go to bed, I had an uneasy feeling. I felt I should double-check the hive early in the morning when it was still dark, before the wasps came back, so I set my alarm for 4:00 AM. But when the alarm rang, I sleepily shut it off and slept for two more hours. It was just light out when I got up, but I decided to check the hive anyway. When I unblocked the hive's front entrance, I quietly observed dozens of my girls take magical slow-motion flight. I watched them ascend to freedom in their circular, carefree flight paths, and it made me smile.

Then, when I looked down into the hive, I was dismayed to discover half a dozen wasps inside. Instead of robbing honey, they were killing bees and robbing sugar water. My hunch to double-check had been a good one. But while I tried to squish the wasps with my leather gloves, scatterbrained me left the lid off for a couple of minutes, and more wasps flew in through the opening. Next thing I knew, dozens of wasps were back in the hive and it was impossible to get rid of them. While I was killing newcomers, more came in. Then, when I put the lid on, I didn't want to lock down the hive, as a couple dozen new wasps were in it. My six-hour lockdown had been useless. Again I had to take off the beekeeping head veil and put on a beekeeping dunce cap.

Sadly, the wasps won the battle against my hive. They killed enough bees to severely weaken it—to the point where it was questionable if the hive could even survive until summer's end. To make matters worse, the militant mites were back in full force, too. I also discovered a disease in the hive called chalkbrood, where the tiny cells get filled up with a spore-forming fungus that kills the bee larvae and makes them look like little pieces of white chalk. Although chalkbrood rarely kills a hive, it can weaken it and reduce honey production. Like a classic battle during the Second World War, my hive was being ambushed. My girls and I were losing each bloody battle, but could we win the war? I went to the library and took out a biography of Sir Winston Churchill and memorized some of his quotes. My favourite was "Success consists of going from failure to failure without loss of enthusiasm."

I was failing at controlling the wasps and had completely screwed up the lockdown by forgetting, I later discovered, to block a third entrance on the *bottom* of the hive. At least I was learning. Everything I was doing, I was doing for the first time, just like when I was a kid. Blocking entrances, concocting poisonous potions, zapping intruders, and wearing a crazy white outfit—it was pretty much the adventures boyhood fantasies are made of, but it was all new to me as an adult.

I drifted back to my childhood. How much fun would those outdoor dinners have been if the electric tennis racket had been invented and my mother had tasked me with defending the roast chicken dinner? Or, better yet, if she had gone inside after the first wasp arrived and come back with our vacuum? What a thrill it would have been for a seven-year-old to save our family from the attacking wasps. My mum, who was from Switzerland, would have proudly exclaimed in German *"Die Wespen sind tot,"* which means "The wasps are dead." I envisioned adulation over my vacuum-hose technique as my granny incredulously sat there enjoying a wasp-free dinner. But, come to think of it, the powerful vacuum nozzle may have been a bit much for a seven-year-old to handle; I probably would have accidentally plugged it with mashed potatoes as I went in for a kill.

SUGAR, SUGAR

After the invasion of the marauding killer wasps, the hive was hurting. The girls were stressed, exhausted, and ill. They needed to be refreshed and re-energized. And nothing provides a pick-me-up like good old-fashioned white processed sugar.

The average North American eats about 130 pounds of sugar a year. This insidious, unhealthy, addictive, sweet substance hides in most processed foods. If that weren't bad enough, we generously spoon even more sugar into our hot beverages and sprinkle it over our meals. My friend Professor Dave Symons—who happens to live in the Beehive State—conducts research on diabetes and cardiovascular disease at the University of Utah School of Medicine. He calls sugar "white death." You must think I am leading up to a claim that honey is a terrific alternative to sugar. Well, sort of. But before I get to that, here is what came as a real shocker to me when I began beekeeping: beekeepers feed their bees the exact same white granulated crystals we all buy from the grocery store. Yup! Kept bees eat processed white sugar just like humans do, and lots of it. Thanks to beekeeping sugar pushers like me, the average hobby hive gobbles up nearly 50 pounds of white sugar every year. This means, factoring in the bee-to-human weight ratio, the average domestic bee eats even more harmful sugar than we do.

Like most people, beekeepers know that white sugar is bad; thus, we eat as much honey as possible. Yet, despite our knowledge of sugar's harmful effects, we feed our bees mountains of sugar. I could see giving my bees a small sweet treat or candy from time to time as a token reward for their hard work. However, it struck me as odd when an experienced apiarist first told me to purchase bags of sugar so big they hardly fit in the trunk of my car for the little junk-food junkies in my hive to gorge on. It seemed counterintuitive—like I was taking them out to dinner at Tim Hortons for doughnuts every night and then expecting them to be healthy and fit.

Here's the rub: northern winters are tough on bees. Through the long cold months, bees burn large amounts of energy while shivering to stay alive; they need to fatten up in the fall if they are going make it until spring, especially if they live on the windy back deck of a float home on a freezing river estuary in Canada. The bees' only source of wild food, nectar, is obviously unavailable in winter. Like humans, they spend most of their winter indoors, but unlike humans, they don't have refrigerators full of food. So in the autumn we load them up with extra sugar calories to prepare them for the long haul.

Contrary to popular belief, bees don't hibernate in winter. To understand why this point is important, here's a mini science lesson. In zoology, birds and mammals (like bears) are categorized as endothermic, or warm-blooded. This means that they maintain their body temperatures internally through metabolizing food for energy. Fish, reptiles, and most insects, on the other hand, are ectothermic, or cold-blooded. These creatures rely on their surrounding environment for heat because their physiology does not allow them to produce heat internally. Thus, they are highly susceptible to external temperature changes. This is why you will see a lizard basking on a rock in the sun on a cool day or a toad sitting in the middle of a paved road on a summer night. They are drawing heat from outside sources. Because evening temperatures are usually cooler, ectotherms are often sluggish at night. Remember how night is the best time to move the hives? Not only are the girls at home, they aren't moving fast.

Because bees are primarily ectothermic, when they are in the hive in the freezing cold, they need a way to generate warmth because they can't create heat within their individual bodies. With the onset of cold weather, bees will congregate in an inner chamber of the hive in a tightly knit cluster that insulates them from the cold. The bees in the centre of the cluster begin shivering their flight muscles (the same way they do right before taking flight), using up valuable stored glucose to create friction. The friction raises the temperature in the very centre of this ball of buzzing bees to about 92 degrees Fahrenheit, warmth the girls gratefully absorb. Coddled in the centre of this live bee sauna is, of course, Her Majesty. The bees on the exterior of the cluster function to insulate the whole ball, and all of the bees are further insulated by their downy "bee hair," called plumose hair. It's easy to see why bees would need a steady source of energy for all that clustering, shivering, and wiggling.

Bees usually create this energy source in the form of honey in late summer. Like squirrels, they pack food away to get ready for cold weather. Bees need the honey they process to sustain the circle of life in the hive year-round. Like us, they eat their stores for energy. Heck, they probably enjoy the taste too. As their first priority, bees squirt tiny amounts of honey into the six-sided wax cells of the comb to provide nutrition for their newly laid eggs. Like us, they put their children first. The remainder of the honey stored, also in six-sided wax cells, is for the rest of the hard-working bees to eat throughout November, December, January, February, and March. Honeycomb does double duty as a bee nursery and food pantry. The amount of honey the bees are able to store in any given summer is their lifeblood. If they have a lot of honey heading into Christmas, they may just live to see Easter.

In spring and summer, beekeepers monitor the hives closely to see if the bees are socking away enough honey in those small cells, and stealing that honey is the selfish motive of beekeepers. In this way we are real-life Pooh bears. But beekeepers have one up on Pooh—knowing full well that the bees are creating winter stores, we trick them into producing more honey than they need. If upon inspection all is well, and we have confirmed there is plenty of honey in the colony, we add a wooden box full of empty frames to the top of the hive. The bees are so busy and focused on gathering nectar that they keep right on filling up those new frames with honey. It's similar to when humans go shopping at Costco: if we know that we have an extensive pantry at home, we stock up on more and more food. Bees tend to follow an ascending pattern while expanding their pantry. Just like humans building condominium high-rises, sometimes the only way to go is up.

So, sadly, all of the bees' hard work gathering pollen and producing more and more honey never really pays off. When the time is right, we beekeepers *steal* their honey. It's the old bait-and-switch routine. Bait them into making more honey, encourage them to store it in an accessible, unsecured area, and then steal it from under their proboscises. When fellow beekeepers wax sanctimonious about the hobby, I sometimes take a bit of umbrage. Don't get me wrong; beekeeping has many merits. It's a wonderful, nature-based learning experience and holds benefits for the environment, not to mention the palate. However, the ugly truth is that this commendable hobby involves ripping off the bees' winter food source right when they need it most, and then replacing it with a $7 bag of processed white sugar. Talk about lunch-bag letdown. As I admit this skulduggery, I confess I am a bit ashamed of being a beekeeper.

The question every beekeeper fortunate enough to harvest the golden elixir must ask is: Have I left my bees enough honey to get them through the winter? In most cases, after blatantly robbing the pantry, the answer is a cold, cruel no. We reap numerous nice little jars of honey to give to family and friends, maybe even jars to sell, but we've left the bees with a big empty upstairs pantry. Because it's getting cold and the nectar in the fields has stopped flowing, the bees can't just run out and get more.

For the honey-pilfering beekeeper, this is where aisle seven at the local Safeway comes in handy. Simply take a left at the Pop-Tarts display, go past the brownie and cake mixes, and you'll find those big 20-pound bags of sugar. It's easy to spot beekeepers in autumn because they are always bumping shopping carts in the sugar aisle as they hoist five or six large sugar bags into the cart for *each* of their hives. Just try putting six of those heavy bags of sugar into a rickety supermarket buggy and step back to watch the cart's frame and wheels buckle under the weight.

To me, feeding bees sugar water seems wrong on so many levels. Granulated sugar mixed into a thick, gooey sugar-water concentration can't be good for any living creature, especially young bees. Think of what happens to kids when they get too much sugar. They get all fidgety and overly energized and then bounce around like crazy. Come to think of it, kids on sugar act exactly like bees.

It might seem less insidious if we fed the bees natural foods with sugar in them, such as bananas or apples, like many parents feed their kids to avoid that sugar high. Feeding them straight processed sugar seems a crime. Still, it's better than having them starve to death. I have heard of purist hobby beekeepers who don't feed their bees sugar syrup, and their bees often don't survive the winter. So I've resisted what my moral compass tells me and implemented the winter sugar-water diet even though it seems insensitive. But what do I know? I am just a bee-ginner, and I don't have the best track record for keeping pets.

Once, for a very short period, I had a cat named Rusty living with me in the float home. He was a feral cat someone gave me when I had rats living in the hull of my houseboat. Although I love animals, I wasn't a very good cat owner because I lived alone and was out of town on business half the time. In the end, Rusty left me.

One day he just up and ran away, eventually, I suspect, dying somewhere in the great outdoors where he longed to be. For the time I had him, I fed Rusty only the finest cat food. I carefully read the labels of all the cat foods at Safeway and chose the brand that would best keep him healthy while he hunted the giant, mean hull rats. I went for the expensive gourmet Little Friskies brand at $2.99 for a small tin, and Rusty loved it. Knowing Rusty was eating well made me feel good. I feel the opposite every time I feed my bees sugar water. The bees need to be fortified to fight the wasps, just like Rusty with the rats, but I don't have the option of choosing premium-brand "healthy" white sugar for the girls.

Ask three different beekeepers the same question about anything bee related, and you'll get three different answers. Sure enough, there's disagreement over the effectiveness of sugar water. The controversy over sugar water is bittersweet. Although feeding it to bees is a hugely popular practice, I recently heard a claim at a bee club meeting that high concentrations of it can harm the bees' immune systems and their ability to fight off diseases. According to my friend Dr. Symons, sugar sure messes with the human immune system. I was also told that beehives fed a white-sugar diet over the winter have a different pH balance than hives that aren't fed sugar. This was on my mind the first time I turned my kitchen into a floating soft-drink bottling plant to prepare winter meals for my bees.

Unlike wasps, bees are not big on eating solid food. A hard grain of sugar would probably choke a small bee to death, so turning sugar into liquid form—which is a cross between art and chemistry—is just one of the skills you'll need if you want to keep bees. Mixing two parts sugar and one part water is the first step in creating the thick, sticky sugar-water concoction. When heating the mixture on the stove, I was taught it's important to almost, but not quite, bring the water to a boil. The maximum temperature to which you can heat the sugar water has something to do with its final viscosity. Turning the stove off just before it boils is key because you are trying to dissolve the sugar, not caramelize it. I don't like making sugar water because I inevitably spill some when I pour it into the Mason jars I use to feed the girls, which makes a sticky mess on the counter and floor. Then my cleaning ladies leave me a nasty note as more ants march in. Plus I don't like the anxiety of having to stand in front of a stove anticipating exactly when to turn the burner off before the mixture reaches a boil. Miss it by five seconds and you have ruined the batch. I get distracted while frying an egg, let alone waiting for the precise moment before a gallon of water boils.

The viscosity of the sugar water is important because it has to flow through the tiny holes you poke in the

Mason jar lid. To feed the bees, you turn the Mason jar upside down on top of the hive. I like to shout out "Come and get it!" as I do. Then you leave the large jar in there for a day or two until it's empty. Theoretically, just like a baby sucks on its bottle, the bees come up to the top of the hive and suck on the tiny holes in the Mason jar lid, extracting the sweet syrup. Theoretically.

The first time I made a batch of sugar water, I was so anxious to feed the girls I didn't let the mixture cool down. I placed the jar of almost-boiled sugar water in the hive. When I went to check the next morning, a dozen bees lay dead next to the Mason jar, their proboscises scorched. Everyone is familiar with those novelty paper noisemakers you get on New Year's Eve—the kind that unroll for seven or eight inches and toot when you blow into them. That is what a bee's sensitive and paper-thin proboscis is like. When the bee is at rest, it is retracted. When she is feeding or drinking, it unfurls into a long tube she uses like a straw. Feeding your bees piping-hot sugar water will burn their proboscises off their tiny triangular heads quicker than you can say "Happy New Year." The girls and I learned the hard way to let the water cool in the fridge, requiring the same patience I had to learn as a kid when my mother was making a tasty bowl of Jell-O.

Once I mastered the art of almost boiling the mixture and had a good cooling method, there were still problems. Half the time, the little holes in the Mason jar lid clogged up with sugar, preventing the flow. I experimented with bigger holes, but that adjustment sometimes resulted in excess pools of sugar water on the hive floor. The bees accidently bathed in the pools while gorging themselves. The sugar stuck to their wings and legs and dried into a solidified white encrustation. The next time I opened my hive, I found sugar-coated, mummified bees. Hmm, possibly a new item I could consider marketing, since I was striking out in the honey production department. All kidding aside, my sloppy, careless bee-feeding techniques and the resulting carnage were disconcerting. At least the girls died happy and full of sweetness.

It took me months of trial and error to get the temperature of the sugar water just right before removing it from the stove, to get the proper mixture and thickness of sugar and water, and to make perfect-sized holes in the Mason jar lids. The recipe had only two ingredients and was simple to remember, yet, despite its simplicity, I still managed to screw it up. As they say, mistakes are what learning is all about.

To continue the learning process and increase my tally of mistakes, I experimented with other feeding techniques and bee dishes. One bee delicacy is called fondant. It is the basic ingredient of many candies, as well as icings for cakes. It's basically a massive soft candy that sits in the hive satiating the bees' daily winter need for sweets. A simple recipe from one of my bee how-to books got me off to a good start. But as I have learned in life and in beekeeping, nothing is simple.

FONDANT --

400 mL water

6 tsp lemon juice

250 mL sugar

Mix the water and lemon juice in a pot, and bring to a simmer on the stove. Slowly add in the sugar, stirring frequently to dissolve. Once the sugar is completely dissolved, heat all of the ingredients on medium high to 234°F on a candy thermometer. Set the mixture aside while it cools down to about 200°F. Using a mixer or your hands, taking care not to burn yourself by perhaps donning some rubber gloves, work the concoction until it achieves a smooth, dough-like consistency. Place the fondant into lightly oiled moulds. Tin pie plates work well.

I always get teaspoons and tablespoons mixed up, and something tells me the exact amount of lemon juice is probably quite important, which may be why my first batch of fondant never made it to a solid state. Down the sink it went. It wasn't really my fault. Whoever named two totally different measurements of volume after spoons both beginning with the letter *t* wasn't really thinking. Why not call them large spoon and small spoon?

For my second batch of fondant, I remembered the teaspoon is the smaller one. With that minor correction, the process looked and felt more promising. I poured the gooey liquid into one of the nine tinfoil pie plates I had bought from the dollar store. When the concoction hadn't hardened after 20 minutes, I put it in the fridge. When it didn't jell after an hour in the fridge, I decided to leave it on the countertop at room temperature overnight.

I woke in the morning and crawled down the 12-foot ladder that leads from my bedroom to the main floor where my kitchen is and discovered the fondant was still not solid. I tentatively touched the surface. It felt super sticky, just like the rolls of flypaper I hang in my living room in the summertime to trap houseflies. Had I created another "meal of death" like the boiling-hot sugar water? The fondant was so sticky I feared that if one of the girls were to touch it, she would be glued there for eternity. I pictured several bees gathered around the fondant, stuck like kids who dare to touch their tongues to a cold metal flagpole. I paused. Even I was not stupid enough to serve them this dud fondant. I decided to store it for a few more days to see if it would eventually harden. After three days, I threw it away.

One of my how-to books had a recipe for something called bee tea. The recipe was far more complicated than any dish I had ever prepared for my friends at dinner parties at the float home. I chuckled when I saw the recipe called not for sugar but for honey. Forget it. Since I had taken over its care, my hive had yielded little to no honey, and the bees weren't getting any of it back. As if that weren't enough, the recipe also strongly suggested using fresh spring water, as opposed to good old water from the faucet. I wasn't sure if the Fraser River could pass for a spring and in the end decided to give up on the whole dumb bee tea idea. It was all a bit too hippy-dippy for me. If tap water was good enough for me, then why wasn't it good enough for my bees? I laughed out loud at the recipe and wondered if I had made a mistake and the tea wasn't for me.

BEE TEA --

    3 cups spring water

    ½ tsp chamomile

    ½ tsp echinacea

    ½ tsp peppermint

    ½ tsp stinging nettle

    ½ tsp yarrow

    ¼ tsp hyssop

    ¼ tsp lemon balm

    ¼ tsp sage

    ¼ tsp thyme

Pinch rue

3 cups cold tap water

1 cup honey

Bring 3 cups of spring water to a boil in a pot. Then take off the stove and add all of the herbs. Steep 10 minutes before straining through a small colander or cheesecloth. Add 3 cups cold water and cool until lukewarm. Add 1 cup of real honey. Stir well.

What a crazy recipe. I lost my Birkenstock sandals in the '70s, and my tie-dyed T-shirts don't fit anymore. I decided I didn't have the right clothes to make this specialty gourmet bee beverage.

The truth of it is, living alone I don't really like cooking for myself, let alone for 50,000 bees. The more I thought about it, the more it seemed that it might be easiest to pick up a six-pack of Coke at Safeway, pop the top, and place it upside down in the hive. But that would be sacrilege.

Consuming more honey and less sugar reaps big health benefits and improves quality of life, for both bees and people. Eating honey has been shown to help people lose weight, as well as reduce body fat and metabolic stress. Honey is also a great fuel for exercise and can improve sleep. Introducing honey into one's daily diet yields dozens of other health benefits. In many ways it is the perfect food.

I suspect that before reading this, you had never really cared about a bee's seasonal diet; however, now that you understand the relationship between processed sugar and beekeeping, you may be a bit pickier about the honey you consume. Relax—if you are buying your honey from a local small producer, you're probably okay. By the time the long cold winter, a time when bees do *not* produce honey, is over, the bees have ingested the sugar and are back out there collecting spring nectar, which will turn into summer's honey. The honey you buy from hobby apiarists is hopefully a melody of nectar from thousands of plants and flowers distilled into nature's perfect food.

The process of introducing sugar to single hives, although not in tune with Mother Nature, is a stopgap measure to bring the hive through the winter. Be aware, though, that commercial honey, available in many large supermarket chains, may not be as good for you as you may think. To accelerate yield, some of the large commercial honey producers—I am talking about producers with thousands of hives—put big barrels of sugar water out in the fields where they keep their hives in spring and summer. For bees in these commercial hives, it doesn't matter how well the nectar is flowing or how many plants are blooming, because the drive-through sugar shack is open 24-7. The big sugar producer in Western Canada is a company called Rogers. There is a joke among local hobby beekeepers that if you feed your bees too much sugar water, you'll harvest Rogers Honey.

When it comes to large-scale commercial honey production, it gets worse. Honey from massive bee farms—with or without the unnatural white-sugar infusion—is often watered down and reconstituted into a hodgepodge of cheaper, less savoury ingredients that act as extenders. Some commercial honey is actually honey-flavoured cornstarch or rice syrup from China. Read the labels carefully. If the contents on the jar have words with five or six syllables, or if you see "high-fructose" anything, take a pass. We derive an inordinate, unhealthy amount of sugar just eating our regular North American processed foods. So when you buy honey, hold out for the real deal. Bee-ware!

Almost every countryside or urban community has a weekend farmers' market. These markets are often in interesting historic parts of town or in the parking lots of community halls. Though numbers of the common western honeybee, *Apis mellifera*, are dwindling and scientists are hustling to determine why, this bee inhabits every corner of North America, and beekeeping is increasingly popular, in part, to offset the losses of wild bees. Thus, it would be difficult to find one of these markets that didn't have a local apiarist sitting at an eight-foot-long folding table with a neatly stacked pyramid of jars of 100 percent natural honey for sale. The jars of golden nectar may cost a bit more than the supermarket variety, but know that your local apiarist is trying to make an honest buck selling you the super-healthy bounty of a well-loved hive.

Remember, as the old Coca-Cola slogan didn't say: Honey—It's the real thing.

CONVENTIONAL WISDOM

On the big weekend of the British Columbia Honey Producers' Association's annual general meeting, conference, and trade show, the weather blew in windy and wet with 60-mile-per-hour winds that tossed the float home around like a cork in a flushing toilet bowl. Trees on the riverbanks toppled into the choppy current. The tempest overturned garbage cans and blue recycling bins. It even knocked out the electric power at the hotel where bee enthusiasts gathered for their long-awaited two-day event. It was brutal. The massive Pacific Ocean can create extremely fast-moving, volatile westerly winds when high- and low-pressure atmospheric pockets battle it out hundreds of miles from British Columbia's shore. These high-speed winds then rage upon the unprotected, open river estuary where I moor my float home and roar through the nearby metropolis of Vancouver. What I can't figure out is how a bee that weighs about one-tenth of a gram can possibly navigate her way through such a powerful gale. It must have been the storm's cold, blustery winds that led one of my honeybees to seek refuge inside my white Volkswagen camper van, which

I happened to be driving to the bee convention that October morning.

With the same Harry Houdini skills the bees used to invade my bee suit, this girl likely found a small opening in the van's sunroof. Or maybe she infiltrated the van's cavernous metal frame by travelling up through the muffler, into the engine compartment, and through the air-conditioning vents until she finally ended up in the sheltered safety of the passenger compartment. All I know is the night before the convention there was no bee inside the van, and all the doors were securely locked with the power windows tightly rolled up. Who knows how she got there, but on that wet and miserable morning after Jeannie and I gathered up the to-go coffee mugs we had filled to the brim, our notebooks, and her 30-pound dog, Tres, and carried them all up the precariously swaying dock ramp to my van, I was shocked to see a bee lying motionless on my grey vinyl dashboard. I chuckled. How ironic was this? Never before had a bee entered my van, and now, on the morning we were going to the provincial bee convention, one lay dead in plain view of the driver's seat. I had half a minute to inspect the corpse while I waited for Jeannie to take Tres for a pee before the long ride into town. When Jeannie returned, and she and the dog had settled into the passenger seat, I pointed to the bee carcass. "Hey, wait a minute," she said. "Maybe it isn't dead. Crank the heat up and let's see what happens."

We sat in the van with the engine running to get some heat while I fiddled with my portable GPS, trying to enter in the coordinates of the hotel, whereupon I discovered it was broken. Uh-oh. When the temperature inside the van finally reached a comfortable level, we began the 15-mile drive to the hotel near Vancouver. The temperature inside the van had risen another five degrees by the time we trekked a few miles down suburban roads and up the on-ramp to the freeway, and this is when we noticed that the ailing bee's lifeblood had started flowing again. At the bee's first slight signs of movement, Jeannie, a natural nurturer, jumped into action, taking the crucial next step to fully revive her.

We carry an emergency jar of honey in both our vehicles at all times. Hey, doesn't everybody? We like to have it handy for when we go to coffee shops and want to use our own sweetener. An emergency honey jar also happens to be useful for providing a boost to anemic stray bugs. My jar was stashed in the storage compartment of the passenger door among some maps and an old plastic ice scraper. Quicker than you could say "glucose," Jeannie had extracted the container, opened the lid, smeared some honey on her index finger, and rubbed it on the dash an inch in front of our frozen insect refugee. It worked like a charm. The lethargic bee smelled the sweetness and slowly advanced her small sickly body. As I drove, Jeannie watched the bee unfold her proboscis, rolling it out into the honey to replenish her depleted system with some of the same golden, life-sustaining syrup that we had stolen from her hive. It was just what she needed, because after a few more miles the bee appeared revived. By that time it was so damn hot in the van that I was sweating. When I complained, Jeannie insisted I leave the heat vents cranked to further aid the bee's recovery.

I skilfully navigated my mobile sauna through heavy, wet weekend traffic, my windshield wipers beating a staccato tempo on double speed and the little bee on the dash cheering me on, but we fell more and more behind schedule. Although I was glad the errant bee had lived, I was still puzzled over how she got there in the first place, and, more importantly, I pondered the philosophical question of *why* she was there. I am a firm believer that everything happens for a reason. Was she a messenger sent by the rest of the bees in the hive? Perhaps she knew it was time for me to learn new apiary techniques and deepen my understanding of what bees need to survive. Was she there to escort me to the convention to ensure I smartened up?

My hive was clearly in rough shape and its future was bleak. Every time I went out to inspect the hive, it had fewer and fewer bees buzzing around the entrance. I'd seen squadrons of wasps attack it, observed hundreds of dead mites on the bottom board, and discovered dozens of dead bees at its doorstep. Not to mention the numerous bees I had accidentally killed while trying to feed and protect them. If I didn't do something soon, my once award-winning hive—the hive that had won second-best-tasting honey at this same convention two years prior—would become a ghost town of three empty wooden boxes.

The terrible state of my hive had embarrassed and concerned me for most of the summer. I hoped the fall apiarists' convention, which cost 300 bucks for each of us to attend, would kick-start me into gear and help me to help my hive turn the corner. Let's face it—I was a beekeeping dud, and I felt guilty. It's one thing to feel bad, but I knew the little bee on the dashboard had to live inside my disastrous hive and endure much worse.

Every day, hundreds of her sisters were dying because of my ineptness. Grim Reaper wasps in yellow-and-black cloaks routinely patrolled the helpless colony, ruthlessly tearing her sister bees to shreds, munching through their heads and legs with chainsaw-like mandibles. The bee on my dash had to live in a real-life bee-rated horror movie. On top of that, she had to wake up every morning inside the equivalent of a dirty apartment with a bare pantry. The hive was full of mites, dead bee body parts, and slimy diseases like nosema, which shows up along with the seasonal wet weather and coats the inside of my hives with brown bee diarrhea. To top it off, neither of the two queen bees I had introduced to the hive that summer had taken. The first one had probably fled the sloppy slum hive, and the second one had a weak, erratic egg-laying pattern. Fessing up to the whole truth of my beekeeper shortcomings, I was away on bike trips so often toward the end of that summer and during the fall that I never did treat the bees for mites regularly or even feed them properly. Heck, I still failed to measure the ratio of water to sugar accurately when I did feed them. In light of this sorry state of affairs and my delinquencies as a beekeeper, the convention couldn't have come at a better time, and maybe the bee on the dashboard was like the Saint Christopher of insects, sent from above to safely deliver me to the great benevolent source of beekeeping knowledge.

Driving in the pouring rain through Richmond, we were finally in the vicinity of the hotel, albeit 20 minutes behind schedule. Without my trusty GPS, I was unsure which road to take next. Heading west on Westminster Highway, I knew I had to turn right at some point soon. I said to Jeannie, "As this little bee is coming to life, I wonder if she knows where we need to go . . . maybe she can tell me where to turn . . . maybe she'll waggle dance us to the hotel." A few blocks after my comment, the little bee miraculously reoriented her torso on the dash 90 degrees to the right. I kid you not—she made a definitive "right turn" gesture just before Number Two Road, which was exactly where I needed to turn right to get to the convention. Divine insect intervention.

When it comes to being on time, and beekeeping in general, why is it that I always show up a day late and a dollar short? We shamefacedly arrived at the hotel just after 9:30 AM. There were already over 200 beekeepers comfortably settled into the convention's first plenary session: a high-quality power point presentation on how global warming is affecting bee colonies. Jeannie and I tiptoed into the room, slouching like juveniles as we slid into a couple of seats together near the back of the room.

When it comes to climate change and bees, you can take an educated guess that any presentation won't be a happy news story. It's hard to distill the tail end of the 45-minute presentation, but here goes. The rise in temperatures is disrupting the sensitive timing of when bees pollinate plants and flowers. Plants are now blooming earlier in the growing season, before the bees have a chance to pollinate them. Hundreds of thousands of years ago, Mother Nature designed the whole pollination–blooming scheme to be a synchronized effort, and global warming is affecting this delicate sequencing. It's not good. The bees may be doomed soon, along with the plants that require their pollination.

I listened to the last of the climate change presentation in the vast ballroom while stealing looks at the sea of beekeepers. Many of the men had beards and wore baseball caps with colourful logos of fertilizer manufacturers emblazoned on them. Waiting in line for a coffee at the 10:30 break, I couldn't help but stare at the hard-working hands of the beekeepers in front of me— the hands of real bee farmers, covered in calluses and small scars and sporting split fingernails. The arthritic evidence recorded upon their gritty knuckles portrayed years and years of stacking heavy beehives onto the backs of old pickup trucks, serious labour that had taken its toll.

I struck up a conversation with a guy just ahead of me who looked like Grizzly Adams. "Excuse me," I said. "I'm just curious. How many hives do you have?" Beekeeper small talk. He was a commercial beekeeper and had come to the meeting all the way from Alberta. He told me he had 8,000 hives. If he noticed my pristine, manicured, light-pink hands, as smooth as a baby's bum, he didn't let on. In a perfunctory manner he asked how many hives I had. Totally embarrassed about my one barely alive hive, I opted for the easy route out and simply replied, "Oh, I'm just a beginner." I continued my small talk, but judging by Grizzly Adams's dismissive lack of interest, I thought it best not to tell him about the cute honeybee that had guided us to the conference that morning. Carrying my cup of coffee to my chair at the back, I felt just a bit marginalized.

Next up was a one-hour presentation on mushrooms. People who study bees are called entomologists; people who study mushrooms are called mycologists. Both terms struck me as good words for a game of Scrabble. People like me who can't keep one beehive alive and who show up half an hour late for beekeeping conferences are called numbskulls.

Somehow the organizers of the convention had arranged for one of the world's leading mycologists to be the guest speaker. Mites, as you now know, can wreak havoc on a hive. Well, this super-brainiac guy from the United States named Paul, who has more degrees than a thermometer and has done 30 years of university research on mushrooms, figures he has cracked the Da Vinci Code on eradicating mites. Could mushrooms be the answer to my hive's problems?

At the beginning of his presentation, Paul explained he had accidently stumbled upon his theory while walking through the woods one day. He stopped to observe some honeybees buzzing around a certain type of mushroom that grows in hollow, rotten logs. During an earlier phase of his research, he had discovered that a rare fungus also found in old-growth forest logs can help fight viruses and diseases, including tuberculosis, smallpox, and bird flu. Strolling through the woods that day he wondered if the honeybee population would see similar health benefits from wood-rotting mushrooms. He went back to his lab, pulled out his microscope, and discovered this: when bees that have visited these special mushrooms return to their hives, they carry microscopic mushroom spores with them. These spores kill the mites in the hive! Needless to say, he was the highly celebrated keynote speaker of the weekend. That evening, he gave another sold-out speech to a convention of mycologists. I guess he found his niche on the bee convention and mushroom meeting touring circuit—a real rock star in the world of mites and moulds.

Although I fell asleep in the middle of Paul's presentation when it got too "researchy" and academic, I enjoyed the beginning when he pulled out a dopey-looking dirt-brown hat made of mushrooms and put it on his head, and I liked the happy ending when he explained how this seminal fungus-based discovery could save all our hives from the mighty mites. I'm not really clear, though, whether we are supposed to plant these magic mushrooms in rotten logs near our hives—that might have been the part where I dozed off. Near the end of the talk, he said something about donating his mite-killing mushroom discovery patent-free to the beekeeping community at large and got another huge enthusiastic round of applause. Then he pulled out another dopey mushroom hat and presented it to the bee conference's master of ceremonies, Jeff. As Jeff put on the hat, Paul explained that the leathery type of mushroom the hat was made of is very rare and can be found only in one town in an extremely remote region of Romania. I ran into Jeff in the men's room later in the afternoon, and he was still wearing the hat. I asked him if I could touch it, and it felt like old damp soft leather.

Other talks that day included a seminar on making mead—honey wine—which was not all that useful in my quest to save my hive. Another one-hour presentation was titled "Bear Fencing," which at least a beginner like me could understand, unlike the next presentation listed in the program, "Germplasm Cryopreservation." I skipped that one and worried that none of the talks were applicable to my personal beekeeping predicament: my dying hive. When I perused the list of topics in the program, I noticed there were no talks, of course, addressing raising bees on a floating home.

Descriptions of the presentations for the two-day convention occupied the centrefold of a slick program: a colourful, thick 32-page booklet full of ads and presenter photos. Conference offerings included the main talks, breakout sessions, and a small trade show. Welcoming messages from important government officials and politicians were splattered across the first five or six pages of the program; however, the top provincial government official in charge of the bee portfolio, which I imagine is not a highly sought-after political position, did not attend the event. Instead, British Columbia's minister of agriculture drafted a missive revealing that the very first honeybee in the province arrived in Victoria's harbour aboard a ship in 1858. He went on to describe the vital role that beekeeping plays in our province's agricultural industry. There was also a welcome note from the mayor of Richmond that was not nearly as interesting as the notes of his provincial counterparts. The aide or public relations flack that writes greetings on behalf of the city's mayor hadn't bothered to research when the first bee settled in Richmond. The feds rounded out the roster of our three levels of government by sending a representative from Health Canada.

I attended her incredibly dry presentation on the National Pesticide Compliance Program. All I learned is that if you are a registered apiarist, and some farmer is spraying a field near your hives with pesticide, you can go up and ask what pesticide the farmer is using, but the farmer is under no obligation to tell you. Jeannie left in the middle of that presentation to take her dog for a walk. When she returned I asked her if our honeybee was still on the dash. Apparently the bee had experienced a full recovery and fled out the window. At least I knew how she got out, because I never did figure out how she got in.

Just before lunch an older gentleman gave a talk on tips for what to do before entering your honey in contests. He had been judging honey competitions at agricultural fairs in Western Canada since the 1940s. I perked up my ears since my Houseboat Honey had won a prize that first summer, but I never realized just how exacting and rigorous the standards were. I wondered if maybe this oldster had actually judged my honey. It was interesting to learn that honey's density and moisture can be measured and defined numerically. I took notes and wrote down that any honey with too much moisture is automatically disqualified from all contests. A small hand-held instrument called a refractometer is used to measure moisture content. The nectar bees gather is about 70 percent water, but the final product, honey, is 18.6 percent or less. It's a universal apiarist rule, and don't ask me why the defining measure is not 18.5 or 18.7 percent. If your honey's moisture content exceeds that magic 18.6 number, you might as well pour it down the sink. Luckily, it will pour easily because it will be too watery.

If you are unfortunate enough to have a dud batch of watery honey, who's to blame? Why, the bees, of course. Just blame it all on them. During the honey-producing process the bees "dry out" the nectar, and one of the ways they do this is by fanning their wings to create airflow around the honeycomb, which helps water evaporate. A water content of over 18.6 percent means the bees were likely not fanning fast enough or long enough—slacking on the job, in other words. Or, maybe, like in my hive, they lacked the sheer workforce numbers or energy and strength to achieve the prized moisture level. During this water-content discussion Jeannie leaned over and whispered in my ear that she planned on buying a refractometer and asked me if I wanted her to get me one for Christmas.

The seasoned honey judge also told us that during the judging process marks are deducted for other reasons like fingerprints on the honey jar, and he suggested wearing white gloves throughout the bottling process. Judges also don't like seeing air bubbles in honey, and there is only one way to get rid of them. You need to get a medical hypodermic needle, hunt the evil bubbles down with the tip, and suck them out of the honey jars one by one. I can't think of anything nerdier than sitting at home the evening before a big honey competition with a magnifying glass in one hand and a syringe in the other, sucking little bubbles out of honey. Though entertaining and peculiar, the judge's information on removing bubbles out of jars wasn't going to save my bees back on the river; I felt a familiar pang of dread and worry.

After lunch, conference attendees got to sample and vote on a dozen jars of honey entered in the People's Choice Award contest. I was blown away by the wide array of tastes, textures, and colours. I savoured each toothpick dollop of honey slowly, closing my eyes as I let the sweet syrup sit on my tongue for 30 or 40 seconds while trying to discern what types of flowers and plants contributed to the flavour composition. Jeannie and I discussed the merits of each sample and really took our time enjoying and comparing. When I was through sampling, I carefully inspected each of the honey jars looking for tiny air bubbles. I couldn't find one single bubble, so I knew what some of my fellow convention attendees had been up to the night before.

Although my convention experience was interesting, I was not really advancing my beekeeping skills—the skills I needed to actually keep my bees healthy and well. It probably didn't help that I don't have the intellectual capacity, scientific background, or attention span to take in two days of intense information on one subject. My mind wanders like a bee buzzing around from flower to flower. Jeannie is more focused and got more out of the presentations than I did. In the final analysis, the pressing question was: Had I, the novice beekeeper, taken in enough practical information to save or at least improve the quality of life in my colony? Not really. I wish I had listened more carefully and not dozed off during the mushroom guy. The convention's large format also made it too hard to ask my naive questions. The hard-working professional beekeepers with thousands of hives intimidated me, and some of the topics were a bit too advanced. I was distracted. To boot, a conference attendee needs sustenance for all that listening, and the lunches they served weren't all that great. I was also disappointed not to get a mushroom hat.

By 4:00 PM on Sunday I was getting really bored. I always feel sorry for the very last presenter at any two-day meeting. It doesn't matter how dynamic or interesting the final speaker is; after two days of sitting on your butt, assimilating hundreds of facts and figures, there comes a point when the brain just goes into "pause" mode. The last presentation was about what types of notes you should take when inspecting your hive. The presenter had invented a codified form of shorthand for recording hive conditions that he was excited to show the group. He claimed his shorthand saved him time and helped him focus on what to look for after he lifted the lid on his bees.

This was finally some relevant information, but long before the note-taking guy even had a chance to wrap up his talk, I was done. My focus was completely shot. So Jeannie and I gathered up our notes, programs, brochures, catalogues, and a wooden bee box Jeannie had bought and walked out into the fresh, moist air and back to the van. Without my faithful "bee GPS," it was a good thing that I knew my way back to the river. The bee was homeless now. I thought about her a lot on that drive. Her chances of survival on her own in a strange town were pretty slim. She needed to live in a hive with her own colony to survive, and although she may well have found another hive within a few miles of the hotel, that hive would likely not have accepted her due to her strange river smell. And the poor thing had no chance of flying all the way back to her home colony almost 15 miles away. What difference would it make anyway? Even if she made it back to my hive, she would be doomed trying to survive in that mite-ridden, filthy combat zone with its shell-shocked queen.

It would have been different if I had learned a lot at the convention and was going back to save the rest of the hive. Her valiant deed would have meant something. She had led me to the source of conventional wisdom and paid the ultimate sacrifice for it. In return I let her down. Truth be told, I didn't even take one single page of notes during the full two days I was there. I napped in my seat more than once and even played on my phone's Scrabble app and surfed Facebook.

My heart filled with guilt and humility, I silently vowed not to give up. I would continue my quest for bee knowledge. I would do it for her. I was determined to become a respectable, legitimate beekeeper like Miriam, Len, Jeannie, and the cap-wearing, bearded bee farmers at the convention. Since the convention experience didn't exactly work out, I promised myself I would seek other sources of bee knowledge. I would actually read an entire book on beekeeping; I would take a beekeeping class; I would join a beekeeping club. Whatever it took. I would not let my bees down again. As my resolve deepened, I looked up through the windshield and noticed the skies had parted and the dark storm clouds had disappeared. Tomorrow was a new day filled with the promise of a healthy, happy hive. With the harsh, ice-cold Canadian winter just around the corner, I knew my girls needed me now more than ever. That little bee on my dash, lost and wandering and forever separated from her hive, would not die in vain.

SWARMING BEE CLUBS

The following spring I attended my first-ever bee club meeting. I have never been much of a club joiner, but I figured it couldn't hurt to go to one bee club meeting to try to expand my beekeeping knowledge. At the same time I hoped to meet some new, interesting people. There must be thousands of bee clubs all over North America meeting in community centres, homes of beekeepers, and, as was the case for me, musty old church basements.

I pulled up to the church parking lot at 6:15 on a warm Tuesday evening, though it may as well have been a Sunday morning just before a service; I was dismayed there wasn't one parking space to be found. I decided to find a place to park on the street. As I drove slowly out of the lot, I scanned the parked cars to glean some insight about the club members' socio-economic traits. It didn't surprise me there was an inordinate number of hybrid cars. In addition, I noticed some classic old pickup trucks and even a 1964 split-windshield vw van representing the tree-hugger fringe. In contrast, there was a sprinkling of late-model bmws and even an Audi suv, which convinced me this crowd had some money.

Clearly none of them would have a hard time ponying up the $30 annual club membership dues.

I spied an open space on a nearby side street and wedged my van in. As I walked back down the tree-lined sidewalk to the church, many other beekeepers were arriving as well. I pushed open the heavy church door, anticipating that I was about to join a healthy club with affluent members.

I had intentionally arrived early for the bee club's 6:30 pre-gathering, called Beeginner's Corner, which I assumed was a safe and casual environment where less advanced beekeepers like me could ask dumb questions without fear of being laughed at. As I descended the worn wooden stairs into the house of worship's basement, I was struck by the ambient sound of voices below. The decibel level of the muffled conversations increased with each downward step.

To the untrained ear, a beehive makes a generic, collective buzzing sound; however, advanced beekeepers can interpret an individual hive's distinctive, ever-changing timbre without even opening the lid and can draw important conclusions. For instance, a certain buzzing sound expresses discontent, while a slight variation of that sound communicates agitation. A sick hive vibrates quite differently than a healthy one, a sound I was sadly growing far too used to. The cadence of an old hive is distinct from that of a young, growing hive.

My float home sits on a scenic bend of the
Fraser River, across from a deserted island.

It was built on an old wooden hull, which was originally a barge
that hauled sawdust up and down the river.

My back deck has just enough room for a hive of honeybees.

The average beehive consists of four or five wooden boxes.

The wooden bee boxes come unassembled and need to be glued and nailed together.

After the hive boxes are assembled and hammered together, they need to be painted.

I like to paint the boxes the bees live in with colourful cartoons.

Different designs I've painted on my hive boxes.

As you may have noticed, many of the chapters in this book are named after my wooden bee boxes.

The three types of honeybees: the top one is the common female worker bee, the middle one is a male drone, and the bottom photo is of a queen bee.

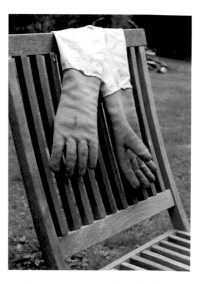

To prevent your hands from getting stung, thick leather gloves are imperative.

One of the most important pieces of beekeeping equipment is the hooded veil.

A sting in the face usually results in uncontrollable, grotesque swelling.

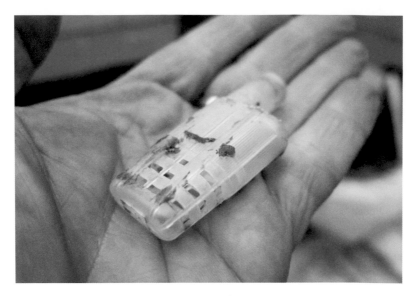

A plastic queen bee cage.

A frame being pulled out of the hive with
the Fraser River in the background.

A typical frame of brood. Note the lighter coloured ring on the outside edge is capped honey, the darker brown oblong–shaped circle in the middle is capped brood.

Bees crawling over capped and uncapped brood.

The beginning of what could become a queen cell.

A frame full of capped honey and bumpy drone brood. Note the three protruding queen cells near the middle bottom of the frame.

Bees drawing out new wax comb.

Large bags full of sugar that
are fed to the bees.

Granulated sugar surrounds
the small opening on top of
a hive's inside cover.

A bee, next to a drop of honey,
that we found on the dashboard
of my vw van as we left to go to
the annual bee convention.

A "field" trip at bee school.

Bees busy at work, building wax comb on a wooden frame.

Removing a frame of bees from the hive, using the metal bee-keeping tool.

A frame where the bees have not yet drawn out their wax comb.

Don't worry...

A giant mural of honeybees.

Winter presents interesting challenges for
bees and float-home beekeepers.

The black dots are dead varroa destructor mites, which after being treated with oxalic acid have fallen off the bees and dropped onto the white corrugated plastic bottom board.

A car battery can be used to power the vaporizer, placed at the bottom of the hive, to combat mites.

The hives securely placed in the back of a truck on their way to the outyard.

A ratchet strap being cinched around a hive to prepare it for transport.

One of the first plants to grow back after a forest
has been denuded is fireweed. Bees love this plant's nectar.

Hives placed behind an electric fence in the outyard.

Honey flowing out of the extractor into a small bowl.

After the honey has been extracted, wax particles and unwanted debris need to be filtered.

Jeannie filling jars of filtered honey.

Small jars of Houseboat Honey ready to be given away to friends.

I've said it before and I'll say it again.

And a productive busy hive loudly and joyfully sings out the abundance within. Descending the flight of stairs to the church basement, I intuited a feeling for the club before entering the room.

The voices drifting up the stairwell embodied a mix of energy, warmth, and interconnectivity. Unceasing, interwoven dialogues were interspersed with plentiful laughter. Based on the number of high verbal pitches, I sensed the crowd skewed a bit more female than male. Furthermore, without clearly hearing any words, I sensed all these people had something in common with my bees: they were industrious, busy, and collaborative. It was the thrum of like-minded people doing something they enjoy: talking about bees. When I finally reached the bottom stair and entered the room, this bee club "hive" was quite loud and very crowded.

In preparation for the meeting, about 150 uncomfortable old black plastic folding chairs had been crammed together so tightly that club members were accidently brushing up against each other as the pre-meeting was called to order. I am fidgety at the best of times and felt sorry for anyone who would have to sit next to me throughout the entire evening's presentations.

Beeginner's Corner turned out to be a talk on the best flowers to plant near your hives. I learned that bees prefer an abundance of white clover and dandelions, but since my float home has no lawn, this was useless information.

When the real meeting finally began at 7:00 PM, there were not nearly enough chairs for the men and women in the room. Over 200 bee club members had arrived, excited to hear the featured guest speaker: a woman from another club who gave a half-hour talk on the life cycle of the wasp. Having recently killed hundreds of wasps, I felt I was an expert on their death cycle. The final stage of a wasp's life, as far as I was concerned, didn't require a half-hour lecture but rather a simple explanation of the differences between an electric tennis racket, a vacuum cleaner nozzle, and a thick pair of leather executioner's gloves. Everybody politely clapped after the wasp lady finished her talk, offering an unintended, partial standing ovation because of the poor club members left standing at the back even though they had arrived on time. I felt sorry for them because the presenters were using a lame PA system, and it must have been hard to hear the presentation. As the meeting dragged on, the basement room became warmer and warmer with no air conditioning or windows to open.

After the wasp review came a power point presentation on queen bee rearing, followed by some remarks from the provincial bee inspector, who explained some of the new regulations regarding the transportation of hives in British Columbia. I didn't plan on moving my hive any time soon, so that talk was not all that interesting to me. But the presentation on rearing queens was fascinating.

I didn't realize that after a queen bee has mated with a drone, she can begin laying eggs within two to three days. I thought back to the dud queen I had bought off that guy on the street and then kept warm in my pocket as I rushed back to the river. She certainly didn't lay many eggs in my hive. Then I got to thinking, "I wonder if she laid some of the eggs in my pants?" Then I got to worrying, "Have I washed those pants?"

One club member then told us about a new book written by a woman who talks to her bees and claims they talk back. Apparently every second chapter was actually written by her bees! I considered whether this writer's bees had stung her in the brain a few too many times. A short discussion ensued on whether the club should invite the author up from Portland, Oregon, to give a talk at the next meeting. I was in favour; I wanted to learn if her bees could proofread too.

Next, we sweated, fanned ourselves, and shifted in our chairs through some boring housekeeping stuff dealing with overdue membership fees and minor changes to the club's constitution, as well as directions on where not to park on the street (where I parked). However, I did learn the club's dues paid for a quarterly newsletter, honorariums for speakers, and the purchase of communal items club members could borrow, such as the bee books and magazines organized into a little library displayed on a folding card table in the corner of the basement.

An expensive stainless steel honey extractor, the size of a washing machine, used to spin honey out of honeycomb was also on the borrow list. You could also borrow nifty gadgets like small moulds used to shape wax into candles. I discovered the club sponsored a number of gatherings and events. There was an annual Christmas party where I assumed that members would display the pretty evergreen tree–shaped candles they created from the moulds. There was another event held on the last Thursday of every month called Bees and Beer, where club members rendezvoused at a pub to drink beer and talk bees. I liked the alliteration of the gathering's title, but I don't like the taste of beer, and it makes me sleepy. However, it was fun to imagine a club member asking the pub's server for "another round of honey ale, please," in a deep, booming voice.

Interrupting my reverie, the club president announced the bee club had hit a milestone with a record number of members. We took a 15-minute break for coffee and homemade banana bread. The coffee break was my chance to connect and communicate with others, just like my bees do while eating. As such, I chose to join two beekeepers, a man and a woman, conversing by the book display. Nibbling on a piece of banana bread, I quietly stood by, waiting for a break in conversation or a cue to introduce myself. They were heavy in discussion about how the club had gotten too big, and it might be time to split off some of the members who lived in the western part of the city into a new club.

"Don't get me wrong," the man said. "It's a great club, but I couldn't find a place to park tonight, and I'm tired of standing. I couldn't hear half of the wasp presentation."

"Yes," the woman agreed, nodding. "I missed a number of the president's announcements. We've simply gotten too big."

Unbeknownst to them, I was hanging on their every word because that was when the club–swarm analogy hit me. Epiphany!

"Wait a minute," I thought. "Everything about tonight's overcrowded, overheated, poorly communicated bee club meeting, including these two members who are considering splitting off and forming another club, screams for comparison. This is exactly what a swarming beehive is all about. Like a healthy hive, it was a thriving club with large numbers of busy, industrious people, most of them women, cooperating and building up the club's common good. And though they were ready to welcome more members like me, the club inadvertently had become too big to sustain any more growth. A split, or swarm, among the membership was a natural and unavoidable outcome, especially in spring, a time of new beginnings for bee clubs and also for beehives."

It is easy to understand why one bee club might morph into two clubs, but why do beehives split in two? To use bee lingo, why do bees *swarm*? There are several reasons. The first is simple: they split because they are healthy. Only healthy, growing hives can split. The same thing can, of course, be said for affluent, well-attended bee clubs. With plentiful members and resources, these clubs are able to split in two, whereas an unhealthy bee club with an empty parking lot and low membership would not likely divide itself. Healthy beehives also split and grow, over and over again, propagating the species, whereas unhealthy hives die out. As the old saying goes, in nature (and hobby clubs) only the strong survive. My hive had never swarmed; it was just shrinking, making me think I should put up a blinking Vacancy sign on the front of the boxes to attract healthy new members.

The second and most compelling reason for a hive to split is overcrowding. Winter conditions can naturally reduce a hive's size from 50,000 bees to between 10,000 and 20,000 bees. If the beekeeper leaves them enough honey in the fall and masters an appropriate sugar-water feeding technique, and if no diseases or mites take hold, the hive survives the cold weather and expands again as the weather warms up. In spring the queen produces more industrious workers, and the colony grows by leaps and bounds until the hive size peaks in late spring with as many as 50,000 worker bees to fan their delicate wings to dry honey and regulate the hive's temperature, guard the colony from wasps, tend to the brood, and forage for pollen and honey. In April or May, with adorable baby bees hatching all over in the bee nursery cells, the hive gets more crowded than our church basement. Thank goodness no one in the bee club brought a crying baby to the meeting.

Overheating, an effect of overcrowding, is another reason hives swarm. Anyone attending the bee club meeting on that warm spring evening could easily relate: club members in that packed church basement felt like Pop-Tarts in a toaster oven. The situation was clear. Yet how do you know when your bees are overheating? The sign that tips beekeepers off is a big, buzzing "bee beard" hanging over the hive's entrance. As the name suggests, a bee beard is a clump of bees that forms the shape of a man's facial whiskers. This formation, however, is about the size of a volleyball. It happens when the house bees push the field bees out of the hive in an attempt to reduce the hive's inner temperature. So, if you come back to your hive one day and discover a formation of bees twice the size of Santa Claus's beard hanging down from the entrance, it means your hive is overheating and a swarm is likely.

The last reason for a hive to swarm has to do with communication or, more accurately, a lack of communication. When groups of people have a falling-out, I have witnessed that it's usually over a communication breakdown. This tendency also holds true for beehives. Remember how bees communicate partially through smell and how the queen and worker bees release pheromones? Those pheromones transmit extremely important messages to the entire hive, and are passed on, in part, when bees share food with one another; when bees are feeding, they are actually communicating. Humans call this "going out for dinner." In the beekeeping world, this communication is known as trophallaxis.

The pheromone released by the queen is like a memo from the head office—she is telling all the bees they must continue foraging, building comb, and tending to the young brood. All the bees know that without the queen they are doomed, so when they stop getting those morale-boosting pheromone memos, they get worried. There is only one queen, and as the hive gets bigger and bigger, not all the workers have access to her and her reassuring smell. When a large percentage of bees stop receiving those sweet pheromone signals, they assume the queen is non-existent and decide it is time to flee the leaderless colony to start a new one of their own.

Compare this scenario to the two overheated club members: their instinct was to flee and form a new bee club. This, of course, oversimplifies the complexity of a beehive swarm, but it does illustrate the point in human terms. You may have already seen a swarm in April or May. People usually freak out when they do. You look out your kitchen window into a tree in the backyard and see 20,000 bees clumped together. If I knew nothing about bees, I would freak out too. Hey, I freak out when one measly bee gets on the inside of my head veil. Often when a homeowner or well-intentioned passerby spots a swarm, that person will call the fire department, police department, or SPCA. Sometimes they call a pest removal company to come and get rid of the terrifying cluster of buzzing, stinging insects. Luckily, all of these agencies usually pass the call on to a local bee club. What is a disturbing bee invasion to one person is a treasure chest of *Apis mellifera* to another.

Most bee clubs maintain a "swarm hotline" to get word out to their club members whenever a swarm is reported. The alerted beekeepers then have an opportunity to capture the swarm, take it home, and put it into a wooden box with empty cell frames to create a free new hive. Some beekeepers I know keep their bee suits in the trunks of their cars, along with a specially designed cardboard nuc box in which they can store a swarm. Their swarm kits each include a butterfly net, branch clippers, a hive smoker, a big white cotton bedsheet, a plastic tub, and the one staple beekeepers can't live without: duct tape. That way if they get a call or text about a swarm, they can proceed directly to the location, scoop it up, and restore tranquility.

My van doesn't, of course, have all the items ardent beekeepers store in the trunks of their cars. I have far too much junk in my trunk already. Maybe that's why I was so woefully unprepared the one and only time I inadvertently stumbled upon a swarm. All Jeannie and I could find to use as a bee suit was an old dirty hoodie in a pile of laundry I was bringing to the dry cleaners. I did have under the van's back seat a Costco cardboard box that originally held six Tetra Paks of almond milk. We could use this as a makeshift nuc box.

The swarm was by the road near some rusty old garbage cans, actually just resting on the ground, not on a branch or post, making it a bit easier to corral the bees, which we accomplished with an old roof shingle we found in one of the garbage cans. Even so, it took a bit of doing to scoop thousands of bees into the flimsy cardboard box with the shingle. After 20 minutes of coaxing and pushing, we finally crammed the little darlings into Almondville. Next, we sealed the top of the box, naively folding the top flaps over one another, and then poked some ventilation holes into its sides. I got stung twice in the process but figured it was a small price to pay for a free new hive. Because we had a vet appointment for Jeannie's sick dog and were running late, we decided to come back to retrieve the box in half an hour. The plan was to race back to the cardboard box after the vet visit, bringing more gear with us in order to complete the safe transfer of the bees in my van back to the float home.

We hid the cardboard box full of bees in a thicket of tall grass, confident we had scored our own treasure chest of *Apis mellifera*. We drove away elated; the promise of a brand new free hive had us feeling giddy. It was too good to be true, and anything too good to be true is . . . usually too good to be true.

When we returned shortly, the bees had all escaped through the thin slits between the top flaps of the box and had continued their quest to their final destination. The box was completely empty, not one bee left. *Jailbreak*.

Swarm day for bees is like Independence Day for a country. It's a seminal moment in a beehive's history—a new colony is formed, a new hive hierarchy is established, and new foraging grounds are claimed. About 60 percent of the bees in the hive just get up and go, and they take their queen along with them. Swarming is like when a country goes through a semi-peaceful revolution and two separate countries emerge.

A few years ago we went cycling for a couple of weeks along the Danube River, through farmlands in Germany and Austria, then into the Slovak Republic. Beehives lined the bucolic route, but we didn't see any evidence of swarms; to be more exact, we didn't see any *bee* swarms. In

1993, however, the Slovak Socialist Republic had swarmed off from Czechoslovakia, forming two independent states. Without getting into a pedantic history lesson—because it makes very little difference to my tale if they took their king or queen with them—the point is that it was a bloodless split, done in peace for the benefit of both states. It was known as the Velvet Revolution, and it is replicated every spring by democratic, peace-loving bees all over the world striving for independence.

Just as separatist movements in countries don't happen overnight, bee swarms require tireless planning, practice, and maybe even some politicking. Try to get 30,000, heck, even three or four of your co-workers or neighbours to do something in unison and you will understand what I mean. Before the bees execute the official swarm, often they will have a few rehearsals. They leave the hive en masse, fly away to some branch or post nearby, stay there for an hour or two, and then return to the original hive.

The key to the success of an actual swarm is the hive's ability to acrobatically escort the queen in the middle of a cloud-like formation. In order for the swarm to survive, the queen must be transported unharmed over the long flight to the location where the bees will establish their new home. Here's where, to use an old hackneyed bee-related cliché, there is a fly in the ointment. The problem is that the queen flies at a reduced rate because she is so out of shape; quite frankly, she's gotten fat because she hasn't flown in ages. She is out of practice. She has not taken flight once or even left the hive since her mating flight when she got knocked up by some drone bees in mid-air years ago. You think I am making this stuff up; I am not.

For years she has been an egg-laying machine, plopping thousands of tomorrow's generation of worker bees into the small six-sided honeycomb cells every day. There is no shortage of meals for her while she toils away. She snacks any time she wants, because all 50,000 bees in the hive want to keep her happy and fed. They feed her constantly, 24 hours a day, and so what does she do? Naturally, she porks out. The problem is that when it is time for Queenie to swarm with the rest of the hive, she is too fat to fly. So, for the weeks leading up to the swarm, the queen goes on a diet to lose weight in order to increase her airworthiness. In preparation for the revolution, she also stops laying eggs. If you inspect your hive and see a slimmed-down queen that is no longer laying eggs, it's a tip that liberation may be well under way. Once the bees have decided on their separatist movement and begun their preparations for the swarm, how do they decide where to swarm? This whole book has been about keeping bees in small wooden boxes. As such, it is easy to forget that boxes are not where bees are supposed to live.

In early spring, while the queen is preparing for the swarm, some of the forager bees, wearing two hats, also act as scout bees. When the foragers are out every day gathering nectar, they scout for possible new hive locations: areas protected from wind, near plants with promising nectar flow, and within a stable and sturdy branch or fork in a tree or, better yet, a hollowed-out hole in a tree. The location's orientation to the sun, its elevation off the ground, and its proximity to water are other crucial factors scout bees note on their reconnaissance missions.

It's natural for bees to swarm back into the trees where they were meant to live in the first place. Just like I don't like to boast about stealing the poor bees' honey and replacing it with white sugar, I also downplay recapturing bees after a swarm and boxing them up. Bees were meant to live in trees, not wooden boxes. I can't help it; I am a child of the 1960s and remember that great movie about freeing Elsa the lion called *Born Free*.

Determining which tree a swarm of bees will relocate to is difficult. *When* the bees will swarm, and by that I mean the season and the time of day, is easier to predict. Bees usually swarm between 11:00 AM and 2:00 PM because their ability to navigate is closely tied to the sun's position in the sky. In ancient times, the nature of a bee swarm—when it swarmed and where it landed—was thought to be an omen or message from the gods.

If you see a swarm in mid-air, you won't forget it for a long time. The swarm begins with thousands of bees per minute exiting the hive's tiny front door and forming a cloud; it looks similar to an ominous storm moving in. Like a macabre nightmare, this small cloud grows bigger and bigger. When the swarm has finally built up to a massive airborne vortex of 20,000 to 30,000 bees, it slowly begins to move in one direction. The creepy swarm doesn't travel too far before it stops to rest. As it moves, an ever-present dull but menacing buzzing pervades the still spring air. Usually the swarm's first rest stop will be only about 50 to 100 yards from the hive. It rests many times along the way to its new permanent home. Resting stops can be staged any time, anywhere, and on anything. A tree limb or fence are where you would expect a swarm to conveniently land, but the swarm's collective reasoning and decision-making process follows no logical path, as far as I can tell. People have spotted resting swarms next to garbage cans, like the one I found, and on fire hydrants, family picnic tables, street signs, car fenders, bikes, and mailboxes. It would be enough to make you pay all your bills online if you woke up one morning, went outside to check your mail, and were greeted by a tight-knit ball of 30,000 swarming bees.

It might seem strange that the swarm moves so much slower than regular bees fly and that the bees have to make so many rest stops. It would seem more natural for the bees to spread out like they usually do when they are foraging and quickly hightail it to their new home. Remember fat Queenie diligently following her weight-loss program to prepare for her big flight? Well, she usually misses the dieting mark. Because the queen is still too out of shape to fly efficiently, the entire swarm escorts her at a reduced speed. She controls the interior of the hive with her pheromones, and she sets the pace for her 30,000 disciples, who clamour to be near her in mid-air so they can feel that pheromone love. When the swarm temporarily lands to rest, the ball of bees compacts as each worker bee manoeuvres and jostles to get closer to Her Majesty.

Now comes the final step of the swarming process, the part I find the most fascinating: the way those 30,000 bees and their plump queen find their way to a new home. If you look closely enough at a resting swarm, you can see several scout bees performing waggle-dance moves on the surface of the tightly packed buzzing bee ball. The waggle dancers are indicating the location of the new home they have found for the secessionist hive and relaying directions to it. With so many scout bees waggling away, the bee ball can look like a mosh pit. Several waggle-dancing scouts perform at once, advising the swarm of a range of potential new hive locations. The more vigorous the waggle dance, the more promising the proposed location.

The queen and the mass of worker bees in the swarm seldom make a snap decision about where to fly to next. A swarm can remain at rest for two or three hours to allow the out-of-shape queen to recharge, while collectively deciding its next move based on the waggling scouts. Sometimes the swarm will deliberately overnight and can even ruminate for days before it gathers itself up and continues on through the air.

This brings to mind a memorable and catchy line in the chorus of the 1961 Brook Benton hit titled "The Boll Weevil Song," which was adapted from an old traditional blues tune first recorded by Lead Belly in 1934. Like the famous line in the song, swarming bees are "just lookin' for a home." Luckily, bees are thought of as better neighbours than the weird-looking, destructive boll weevil with its long snout and penchant for eating the flowers and buds of the cotton plant.

Once the swarming colony is settled into its new branch, tree fork, or trunk cavity, the cycle of life continues. As for the old hive, the one they left behind that still houses 40 percent of the bees, well, it now needs to create its own new queen in order to continue on.

If you ever see a bee swarm, don't be frightened. Simply give it a wide berth and leave it alone, then call the friendly experts at your local bee club. Or, if you love *Born Free*, simply continue on your way and let the bees continue on theirs. Above all, don't touch or try to contain a swarm, unless you are a knowledgeable beekeeper with the appropriate capture paraphernalia in your trunk. Messing with a swarm could be messy and backfire. The same can be said if you stumble on a bee club meeting in a church basement. Just leave them alone.

I walked back to my van after the bee club meeting, thinking it was pretty clear a club swarm was inevitable. Should I stay with the old club or join the clump of people flying to a new location? If I were to join the swarm, when would we swarm, and to where? Would we do a trial swarm by having dinner together one night at a restaurant near the old church?

On the way back to the float home I passed three churches that seemed larger and more suitable than the one I had just left. I also passed a large modern community centre that I knew had a sizable gymnasium. The centre could easily fit a growing bee club. So I began daydreaming about the next bee club meeting and how, during a coffee break, I would communicate the location of the community centre to my fellow club members. It was obvious. While shoving crumbs of banana bread into their mouths, I would waggle around in a figure-eight pattern with an axis that pointed toward the community centre while frantically flapping my arms up and down.

BEE SCHOOL

The beehive that Miriam and Len had bravely left in my care didn't come with an owner's manual. Even if it had, you know by now that I probably wouldn't have taken the time to read it. Everyone in my circle of beekeeping friends and family had 100 times the knowledge of bees I had, so it was no fun always asking them dumb questions like the ones I asked Miriam: "Just how many eggs does a queen lay each day?" Or, "How often should I check for mites?" Or, my most frequent inquiry, "When will I be able to finally harvest some honey like we did in the halcyon days when you first dropped off the hive?" It was no fun having zero jars of honey to hand out, especially after I had boasted about how great my hive was to so many people. I had established this faux reputation of being a real beekeeper. It was definitely no fun waking up in the morning and walking out onto the back deck to check my hive only to sweep dozens of dead bees off the deck into the river.

My ineptness was starting to get me down, and I needed to re-establish my reputation among family and friends as a competent beekeeper. So I signed up for a series of apiarist classes. The cool thing about this beekeeping course was that if I could pass a final exam, I would receive a frameable piece of paper from the British Columbia Honey Producers' Association proclaiming that I was a certified beekeeper. The provincial Ministry of Agriculture helped to develop the course curriculum, so it was legitimate. Plus the certificate had a really cool embossed gold stamp. Jeannie had completed a similar course a few years earlier, but at the time the course hadn't offered the provincial certificate. I loved the idea of being a provincially certified beekeeper when she wasn't, even if my one and only hive was dying while her hives were producing upwards of 300 pounds of honey. Miriam and Len had certificates, but I'm not sure if they required writing a test. At least if I got accredited, I'd have bragging rights. The course cost around $250—a small price to pay to save my hive, while bolstering my ego and restoring a few shreds of self-esteem.

In actuality, other than looking academic and official, the certificate serves no real purpose. It's not as if you have to show a certificate to someone to buy a hive or sell a jar of honey. And it's not as if the bees care either. But heck, I never went to university, and the thought of framing the certificate and hanging it in the float home near my beehive appealed to me. The only problem was the course came with a 205-page textbook I was supposed to read, and I had to actually attend 24 hours of classroom instruction over 12 weeks.

I did lousy at biology in high school and still carried some unhealthy emotional baggage from my Grade 10 experience with a certain science teacher who shall go unnamed. He gave us a 10-point quiz that required us to memorize the parts of a human eyeball. We had to notate the terms on a crudely drawn black-and-white eyeball diagram. I got the parts all mixed up and received a 0 out of 10 on the quiz, even after I studied for 45 minutes the night before. When the insensitive teacher—who had obviously chosen the wrong profession—handed back the tests, he made fun of me in front of all the other students. Of course, they laughed, all 30 of them. Ever since that day, I have had a neurotic aversion to written exams, especially when they have to do with biology.

I was pleased to discover the beekeeping class had fewer students than my high school classes, only 14 or 15, and they were all adult beginners like me. Some had hives and some were considering getting a hive. At the start of the first class the teacher handed out the syllabus, which outlined the topics we would be required to learn over the next three months in order to pass the course. They were: the basics of honeybee biology, municipal bylaws, proper equipment selection, bee acquisition, identifying hive diseases and assessing overall colony health, honey production, apitherapy and the products of the hive, honey harvest and extraction, winter preparation, bee life cycles, and capturing swarms.

During the first class session I sat in the middle of the front row reviewing each week's lesson plan while sipping a disappointing lukewarm coffee out of a paper cup. The curriculum looked somewhat interesting. The only catch I noticed was that in order to actually get the certificate, I had to pass the final exam by getting at least half of the questions right. The teacher explained that all 24 hours of classroom instruction and all reading assignments boiled down to this 50-question final test, with each question worth two points. The textbook on which the exam was based had some nice colour pictures, but I later discovered the writing was torturously dry. Still, I felt optimistic about my chances for success. On the inside wall of my float home, the wall nearest to my hive, I had hung a calendar, which my bank had sent me in the mail, to track when I fed my bees, when I last checked for mites, and other important beekeeping journal entries. When I got home I walked right over to the calendar and marked the date of the upcoming final exam with a giant red X. If I were successful, in 12 weeks I would display my beekeeping certificate on the wall right next to my calendar.

But the truth was I hadn't been examined in a classroom setting for over 40 years. I teach a university-level international sports marketing course part-time in

Vancouver and Vienna, and thus I am used to handing out exams and marking them, not taking them. So as the bee exam got closer and closer, I started to mildly wig out. Latent neurotic negative feelings of possible exam failure and the humiliation that accompanies ignominious academic defeat haunted me. I began having a tough time falling asleep at night as I tried to memorize facts presented in class that I thought would be on the test, but since I hadn't taken any notes, it was rather hard to memorize them. It didn't help that I had missed about a third of the classes as well. And I never completed reading the boring textbook. This was looking grim.

I had trouble committing to memory fundamental facts, such as the differences between venom, propolis, larvae, and royal jelly, or the best way to treat your hives if they have American foulbrood disease. I knew the exam would be asking me these and dozens of even harder questions, like the distance apart that each wooden comb frame should be from the other frames in the hive boxes. Was it half an inch or three-quarters of an inch? I tried to memorize all the major parts of a bee's body, like the head, abdomen, and thorax, but there were 15 smaller subparts that I just couldn't get straight. Plus, learning biological parts of any kind brought that eye diagram swimming back into my brain, along with the word *zero* in angry red ink. In one class session we were taught that the inventor of the modern hive was some guy named Langstroth, which I already knew, and we had to remember that he patented the hive design in 1852 . . . or was it 1853? I tossed and turned each night in bed, my mind a jumbled mess of bee body parts, disease names, biological theories, city bylaws, hive construction measurements, historical facts, and other easily forgettable bee minutiae. Bee exam insomnia translated to daytime fatigue.

With the exam less than two weeks away, my anxiety worsened and began to affect my day-to-day routine. One day I paid for my groceries at the supermarket and forgot to take the two bags of food out of the store. *True or false: worker bees must have protein, carbohydrates, and water to raise brood.* Later that same week I missed the highway turn to the float home. *If a queenless colony has only capped brood and no eggs, does that mean the queen was present and laying eggs eight days ago? Is it true that the medicine fumagillin is used to treat the disease caused by* Nosema apis? I was so nervous I was stumbling over the most rudimentary questions. *How many legs does a honeybee have? How many wings does a honeybee have?* I knew the answers were four and six but was not sure in what order. I feared this final beekeeping exam was shaping up to be a disastrous repeat of my Grade 10 biology test. During the second-to-last beekeeping class, one week before the big exam, I looked around at the rest of the men and women in the room. If I failed the exam and they all passed, would they laugh at me? Do we ever really outgrow our adolescent cruelty?

I didn't know the answer to that question, but I wasn't about to chance it. I hate to admit what I am about to tell you, but compelling creative writing can only come from total and direct honesty. Here it comes: I cheated on the bee exam. I am not proud of this, and I can come up with numerous excuses for why I cheated, but in the end, I admit it right here, right now: I cheated.

I used an old trick I learned in high school that I call "shorthanding." I was quite certain that listing a bee's body parts was going to be a question on the test, so I took the first letter or two of the 15 body parts listed in chapter 11 of the textbook and wrote them out on my hand in blue pen. I'm not talking about thick felt-pen markings, just small subtle letters inscribed with fine-point blue ink that only I could see written secretly on the fatty part of my hand under my thumb. Teeny tiny letters that only one person in the whole world knew about. Based on my limited experience of cheating in high school, I remembered that the fleshy part of the hand, large and flat enough to accommodate 30 or 40 characters, naturally faced down on the desk, hidden as you sit; thus, it's easy to secretly turn your hand over to read when the teacher isn't watching.

I know what you are thinking: "When you cheat on beekeeping exams, you only cheat yourself." But I figured I wouldn't need to know all these microscopic bee body parts in the future. All I needed to do was remember where I left the course textbook, and that way if the need ever arose, like for instance if I needed to take one of my bees to the dentist because it had a sore mandibular gland, I could refer to the book to learn where that gland was.

Perhaps you're getting a bit judgmental at this point and thinking I should not have taken the shorthand shortcut. You might feel I should have just hunkered down and memorized all those bee parts. Okay. Bee my guest and here they are: antenna cleaner, proboscis, mandible, mandibular gland, antennae, hypopharyngeal gland, compound eye, ocelli, forewing, hindwing, wing hooks, Nasonov gland, stinger, wax glands, pollen baskets.

Boring.

After reviewing that exhaustive list only once, I am sure you will agree with me that one could save a lot of time and energy with 15 helpful, hidden, one- or two-letter hand hints: AC, P, M, MG, A, HG, CE, O, FW, HW, WH, NG, S, WG, PB. When I snuck a glance under my thumb, each one of those letters ignited some galvanic nerve memory response deep in my brain that would recall the correct word. It wasn't really cheating; it was more of a little helping hand.

On my other hand, I wrote a cryptic series of words, numbers, and letters to help me answer other predictable questions. These facts, on the other hand, were important to know as a beekeeper, and so rest assured that I have since gone on to commit them to memory. They dealt with the number of days it takes after an egg has been laid for the egg to develop into one of the three types of bees that inhabit a hive. This is practical information that can lead to clues about when the queen was last laying eggs and when newborn bees will be arriving. A queen bee takes 16 days to develop from egg to bee, a worker bee takes 21 days, and a drone bee takes 24 days. In shorthand written under my thumb, I noted the following code: "Q-16, W-21, D-24." In the bit of room left over on that hand, I wrote the word "India," which is where the North American honeybee originated, and also the letters "AHB" for African hive beetle, which I had learned was a new pest threatening hives in North America. Thumbs-down on those beetles invading my hive; thumbs-up on getting a mark for knowing its name.

I'll admit I felt a bit stupid, a tad guilty, and slightly foolish sitting at my kitchen table writing tiny letters under my thumbs in the hours before the final exam. When I ran out of room on both hands, I stopped writing, grabbed the keys to my van, and left to take the test. Driving to bee school that night, I worried that if I ever got caught and ended up in Bee Exam Court, it would be considered first-degree cheating. Just like first-degree murder, first-degree cheating is premeditated. But I figured I had a low risk of getting busted. The lady invigilating the bee exam probably wouldn't be paying that much attention. Besides, it's not as if she really cared, and anyway what kind of dopey, lazy low-life in their late 50s would cheat on an exam for a hobby? It was not like this was an important medical school surgical exam where my studying shortcut would harm a patient. It was a victimless crime. Um. Well. That's not entirely true. There were 50,000 potential victims.

While the pangs of guilt plagued me, I was actually quite proud of my two fleshy cheat sheets and gloated to myself over how well and neatly I'd prepared them to address dozens of anticipated exam questions. I was confident my inky codes would be well worth the time and effort and pay off with a passing grade. The gold-embossed certificate conjured visions of those stately framed degrees hanging in the offices of high-powered lawyers.

But as I pulled into the school lot and parked the van, I began to sweat. The real test was just minutes away. I worried the sweat would smudge the tiny letters on my hands. Upon entering the classroom, instead of sitting in the front row like I usually did, I sat in the very back row, not that I was being obvious or anything. I was now sweating profusely, realizing I may have inadvertently made a wrong move that would bring unwanted attention from the teacher. When she glanced over at me, I nodded and then, like a good test taker, I neatly placed my binder, textbook, and notepad on the floor beneath my seat. When I hand out an exam in my marketing class, I like it when the students clear their desks because it demonstrates respect, seriousness, and an honest, fair approach to the task of writing an exam. The teacher walked down each row of tables handing out the test papers. I smiled and clenched my fists as she placed my exam on the desk in front of me. When she was gone, I neatly wrote my name on the front page, blood pressure surging like I'd just been jabbed with an EpiPen. I turned to the second page where the very first question asked me to list the body parts of the honeybee. Bingo!

Cheating, like most of life's vices, can be addictive. After flawlessly listing all 15 bee body parts and then choosing responses to the multiple-choice section from the answers recorded on my hands, I had to move on to the rest of the exam without assistance. I did my best ticking off the appropriate true or false boxes and then faced the fill-in-the-blank questions. After 50 minutes of deep thinking, I was drained—my brain was quite empty—and yet eight questions remained unanswered.

They were all true or false questions on the last page, and I didn't have a clue how to answer them. I hadn't written the answers to these questions on my hand, and even if I had, I'd sweated so much for the past hour that the finely drawn tiny letters had dissolved into blue smudges. I read each of those eight questions again and again and again, panicking, my vision of the gold-embossed provincial certificate dissolving as the clock ticked toward the end of the exam period. I desperately needed a lifeline.

T / F Formic acid treatment of hives controls both varroa and tracheal mites.

T / F All medications need to be removed from the hive prior to the honey flow.

T / F Locating hives in a straight line will promote drifting.

T / F Chalkbrood mummies are a source of chalkbrood disease.

T / F The Bee Act of British Columbia requires inspection prior to selling your bees.

T / F An ionizer is used to rid the hive of nosema.

T / F It is best to avoid facing your colonies toward the south.

T / F Feeding bees purchased pollen can infect the colony with American foulbrood.

I hate to admit it, but, racking my brain, desperately trying to figure out the answers, I cheated some more. This cheating was totally unplanned and more a matter of taking advantage of an opportunity that presented itself. A gift. It was spontaneous, and as such, if the British Columbia Honey Producers' Association ever pressed charges against me and my transgressions ended up in Bee Exam Court, the judge would surely pass a more lenient sentence, perhaps letting me keep the certificate but marring the gold embossment by drawing the letter *C*, for *cheater*, across it with a red Sharpie.

The classroom we wrote the exam in was lined up with eight-foot tables. I sat at the back at one of those tables with two other students. With three people per table, we were so closely seated that it was easy to glance over at each other's test papers. There was a middle-aged Asian lady with long black hair sitting next to me. She was the best-dressed person in the class with tortoise-shell bifocals and a smart brown blazer with an embroidered, classy red silk shirt under it. She looked smarter than me—come to think of it, everyone in the class looked smarter than me—but she looked much smarter than me, so I casually spied on some of the answers on the last page of her test sheet. Another victimless crime.

We were given a full hour to write the test. Most of the others left when they were done after 35 or 40 minutes. I used every last grain of sand in the hourglass to complete mine, agonizingly reviewing each question over and over again until the bitter end. Then, right after the time ran out, what did I do? Why, I destroyed the evidence, of course. I hurried straight to the bathroom and, like a surgeon preparing for an operation, rigorously scrubbed both of my hands, massaging the ink out of the fleshy parts under my two thumbs. A profound sense of relief came over me as I watched the blue-tinted water swirl clockwise down the sink's drain. I didn't stop until my hands were restored to their original pristine pink colour. I was glad I wasn't one of those rustic bee farmers from the convention because the ink probably would have firmly lodged in the cracks and calluses of hard-working hands, allowing the evidence to linger for weeks. But for me, two minutes of the most thorough handwashing of my entire life felt nothing less than cathartic. I left the washroom an innocent man.

Three weeks later, an envelope from the British Columbia Honey Producers' Association arrived in my mailbox. I didn't want to get my hopes up too high, but it wasn't a standard-size envelope; it was big enough to hold a beautiful 8½-by-11-inch provincial beekeeper certificate. I was breathless as I opened it. There is no way I failed this test, no way. I knew a third of the answers from attending some of the classes, another third were written on my hands, and the rest I filled in from copying my neighbour's test. As I ripped the envelope open, I discovered three separate documents. The first document was an evaluation form for the course. I briefly reviewed the typical class survey questions on course content, handouts, and instructor effectiveness. I paused when I got to the box for comments and decided not to point out that I liked the final exam seating arrangement of three people per eight-foot table. I crumpled up the evaluation form and threw it in the garbage. With trepidation, I reached for the second document that I could see was three stapled pages—my marked final exam. I almost dropped to my knees in elation when I saw 84 out of 100 marked in bright red letters at the top of the test. The teacher had scrawled, "Good job, Dave!" Her compliment brought me not a twinge of remorse; I can't believe how low I stooped. Egad, what a cad.

The B mark was fantastic. It really lifted my spirits. But I was most excited about the final piece of thick parchment paper—my *official* beekeeper certificate. It was a beautiful proclamation with an ornate gold-leaf border and my name written in the middle in fancy calligraphy announcing, "This is to certify that DAVE DOROGHY has successfully completed Introductory Beekeeping," and signed by the president of the British Columbia Honey Producers' Association and the course instructor. No one could take the accomplishment of passing the test with flying colours away from me now, and I had the certificate to prove it. My old high school science teacher could shove that eyeball diagram in a certain biological orifice that I would also be tested on later in Grade 10. I was at the top of the class in bee biology, baby!

Postscript: Although I feel no twinges of guilt or remorse, others may want to confront my transgressions.

First, I admit to cheating on the test. I am hoping there is some kind of statute of limitations, and the British Columbia Honey Producers' Association isn't going to come knocking on the float-home door, demanding the certificate back. Just to be on the safe side, if some authoritative-looking person in a beekeeper's suit comes knocking on my front door, I just won't open it. Or maybe I can scare them off with my electric tennis racket fly swatter. Hey, it works with the wasps. Just to be doubly safe, I took the certificate to the library and made a high-quality colour photocopy of it. If they revoke the certificate, I will be sure they get the copy, and I'll keep the gold-embossed version.

Second, I never told Len or Miriam or Jeannie that I cheated on the final exam. They'll find out about it by reading this. I figure the points I lose with them will be regained when they realize I wrote a book about bees. It'll be a wash.

Now, does a bee have four legs or six? Who cares? Google it. I got a Bee plus in the course, and I'm a certified beekeeper, and that's all that matters.

FLY UNITED

When I was 12 years old, novelty posters were all the rage. I mean those giant humorous posters you'd hang on your bedroom wall, much to your mother's chagrin. I'd save up my weekly allowance and send away for them from mail-order ads in the backs of the old Archie, Charlie Brown, Tarzan, and Richie Rich comic books I collected. These were posters with adolescent, sexually charged double entendres and colourful graphics. One memorable poster depicted two cartoon ducks copulating in mid-air with the headline "Fly United." Very popular in 1969, this poster was by far my favourite.

Around the same time that I was tacking that poster above my bed, I was hiding *Playboy* magazines under my bed. The May 1968 issue—I remember the month because I had the worst crush on Miss May—had an article about a certain unusual "swinging" group of grown-ups having sex in the bathrooms of commercial airplanes. The article went on to explain that once you have "done it" in a plane, you join an elite clan of oversexed adults called the Mile-High Club. At 12, I had never even been in a plane. Since then, I have been in many planes, but I still don't know a single person who has actually had sex in an airplane washroom, and I can't imagine that it would be much fun due to the cramped space and unpleasant odour. I was a skeptical kid, so I never really believed the cartoon birds had sexual intercourse in mid-air either. Since my adolescence, I hadn't given much more thought to mating in the sky. Until I became a beekeeper, that is.

One particularly riveting class at bee school taught me that bees really do have sex in mid-air. It's not a cartoon gag; it's not a fantasy. It's not the scandalous, semi-fictitious subject matter of an erotic, sensational 1,200-word article buried behind the iconic three-page gatefold topless model in an old *Playboy*. It's just the plain biological truth and a scientific fact. That's where bees do it: in mid-air. When you think about it, it makes sense for bees to "fly united" in the friendly skies, since they are less vulnerable to predators when they are flying.

If I thought as a teen that *Playboy* was racy and explicit, the magazine had nothing on the beekeeping textbook handed out as part of my certified apiarist course. Although the colour pictures of bees held much less appeal than Miss May, the salacious description of what the virgin queen bee does more than made up for it. The way bees mate in mid-air is pretty amazing and beyond anything I ever fantasized about as a young boy.

Anyone reading beyond this point must be 18 years old or older. Erotic bee sex spoiler alert!

As you know, the queen bee is a superpowered egg-laying goddess, efficiently creating up to 2,000 perfect little eggs a day in the hive for months at a time. However, she can't do it alone. In order for an egg to eventually create a baby bee, it takes two. Now, as the song goes, "let me tell you about the birds and the bees and the flowers and the trees and the moon up above and a thing called love . . ."

It actually takes more than two when it comes to the queen bee's mating ritual. On her mating flights, the queen will mate with between 15 and 20 different male drones. The articles in the faded and dog-eared pages of my old *Playboy* started looking pretty tame once I considered the nature of Miss Queenie.

So where and how does the queen bee get her action? In a typical hive, tens of thousands of industrious female worker bees, hundreds of testosterone-filled male drone bees, and one fertile queen all cohabit under one roof. Compared with the bustling activity of the female worker bees gathering pollen, building the comb, processing the honey, nursing the young, and warding off intruders, the drones seem pretty lazy—they just eat and mate. But the drones aren't just loafing around. They can't collect pollen or nectar because of the way their bodies are designed. Drones don't even have stingers—they are all bark and no bite. Talk about a good excuse not to have to work! Besides, they must save all of their energy for lovemaking. They live, essentially, to have sex with a queen during mating season, and for the time leading up to that, they simply lounge around in the hive all day, surrounded by females working hard to support them.

In many ways drones are a lot like the controversial founder of *Playboy*, Hugh Hefner. According to articles I read in his magazine as a kid, Hefner, wearing his black silk housecoat with the *Playboy* logo embroidered in gold on the front chest pocket, lounged around in his mansion all day surrounded by beautiful women. The women supported Hef by working hard on photo shoots that made him lots of money. Unlike Hefner, who lived to his 90s, drones only live for up to four months. And unlike the skewed power dynamic of the Playboy Mansion, in a hive, the women are clearly in charge.

When mating season is over and the queen is successfully pregnant and laying eggs, the fat, unpopular drones are a drain on the hive, and thus the ruthless female worker bees, which live for about 40 to 50 days in summer, boot them out. Usually the displaced drones loiter around the front of the hive like punks ejected from a popular nightclub. Without access to the golden honey, they eventually starve to death.

I know what you may be thinking now: with all those local homegrown drones looking to get some action and only one queen in the hive, what about insect incest? What's to stop a virgin queen from mating with one of her own brothers? Well, it's possible, but unlikely. You see, when the queen takes off for a sex-crazed night on the town with a bunch of young, virile drone bees, she finds her very own Mile-High Club to join far, far away from the hive and from her children. These clubs are located about 200 to 300 feet above the forest floor. Queen bees may travel several miles to reach one of these special drone congregation areas. Massive clouds of drones, some 25,000 bees from as many as 200 different colonies, gather at the exact same spot every year to mate. Finding her "hookups" in this testosterone-filled cloud of out-of-town drone bees is how a virgin queen can guarantee that she will receive the genetic diversity needed to ensure the ongoing survival of her colony. Somehow she knows that by flying to drone congregation areas farther away from her colony, she will avoid inbreeding. And this, of course, also reduces the embarrassing possibility of running into one of her brothers. "Hey, Sis! What are you doing here?"

When I think of drone bees, I think of men in their early 20s who like to gather together in singles bars and ogle the pretty young girls flying by, hoping to meet one. With thousands of drones vying for the attention of one female queen bee, the competition is fierce. The only place they can score is at one of these drone congregation parties, which occur annually at a very specific time and place. But just how do the drones determine the exact spot to host these swinging parties? It has something to do with the proximity to nearby trees, the sun's position in the sky, and the direction of the prevailing winds. Why the following year's crop of drones goes back to the same spot, at the same time, is anyone's guess. People who track these types of things, people I call "strange insect voyeurs," have documented drone bees congregating every year in the exact same spot for over a dozen years. And the queens, who have very little experience in navigating the great outdoors because they are so young they have logged little or no flight time, also mysteriously manage to find these singles clubs. Since drones and queens intrinsically know how to find these sex rendezvous spots, perhaps it's written in their DNA. Observation reveals that other animals must have similar mating DNA imprints.

Take salmon and their spawning habits. When it comes time to lay their eggs, the salmon also know exactly where to go. Living on the mouth of a river, I know a thing or two about these glamour fish and their amazing life cycles. There is a big salmon-processing plant about a third of a mile away from where I moor my float home, where gillnetter boats drop off tons of fish during the annual salmon runs. Millions of salmon are born each year on tributaries of the Fraser River, way up in the heart of British Columbia. After they are born, they travel thousands of miles—as far afield as Alaska and Japan—but when it is time for them to spawn and lay their eggs, they return from overseas four years later to drop their fertilized eggs in the exact spot where they were born. The process takes the fish right underneath my float home and my beehive twice, once on the way out to the ocean and once years later on the way back.

Then there is the mating ritual of the emperor penguins, in which males and females begin their mating dances miles apart from one another in temperatures a gazillion degrees below zero. Each black-and-white tuxedo-clad bird waddles about 50 miles from locations around the South Pole to inland Antarctica in order to meet at a specific breeding colony. Again they know the exact spot to rendezvous. Then they stand in a crowd and the males sound a loud bugle call for the females, who can recognize their mates' voices. After that, they take a long stroll around the group, bow deeply to one another, nuzzle, and make strange noises before mating.

How do these bees, fish, birds, and myriad other animal species know where to go to find their lovers? It's a mystery. For now let's just say that Mother Nature has done an amazing job here on earth giving all living creatures, including people, an instinct for finding their mates. In Vancouver in the late '70s, clubs like Oil Can Harry's, Pharaoh's Retreat, the Raspberry Patch, and, my favourite, Misty's, were all the rage. These rendezvous spots had an abundance of queens flying around inside, all dolled up with amazing hairdos, sweet perfume, high-heeled shoes, and killer dresses. My animal DNA imprint forced me to dress up and go out to find them. I increased my chances of mating by wearing the hippest robin's egg blue leisure suit with bell-bottom pants. I stylishly tucked my shirt collar on the outside of my jacket and carefully left the top three shirt buttons undone to expose my limited chest hair. I then splashed on copious amounts of Old Spice aftershave lotion. Did I mention the platform shoes? I'm six feet one. If you added two inches, I became one of the biggest, brightest drones on the dance floor. The drone bee in a hive is bigger than the females, and the bigger the drone, the stronger and more suitable a mate. By making myself big, like the drone, I had the edge on attracting disco queens. Although, more times than not, in some weird kind of Darwinian evolution of the species selection process, I would usually get turned down by most of the women I asked to dance.

Every once in a while, though, a queen would agree to trip the light fantastic with me—unfortunately, it seems, more out of curiosity than a desire to mate. Once we made it out onto the dance floor, however, I delivered! Look out! I could waggle dance with the best of them. Turn the knob on the professional sound system amplifier up to 11; throw the latest hit from the Bee Gees, ABBA, KC and The Sunshine Band, or Barry White on the turntable; and set me loose on the dance floor to do what I was programmed to do: attract a suitable mate. I had this certain dance move women found irresistible. I'd bend my knees slightly, swing my hips from side to side, and, while wiggling all 10 of my fingers to the music, I'd gyrate my arms horizontally, raising my hands from the akimbo position on my hips up to my brow. Every move was executed in perfect syncopated rhythm with the pulsating disco beat. I called this surefire dance move the Shy Tuna. Doing the Shy Tuna under the disco ball, sporting fashionable long sideburns, a Burt Reynolds–style moustache, and a heavy gold necklace often led to instant primal attraction. Well, it did once. Well . . . almost once, until her roommates came home unexpectedly and ruined the whole thing.

On the female side, comparisons between bee mating and human dating rituals continue. Shortly after young women reach sexual maturity, many become interested in boys and hope to begin dating. This is when they pester their parents to allow them to fly the nest to go hang at the mall or go to "a friend's house" (secret code for party). This dating interest also happens quickly for the queen bee, even more so as she hits sexual maturity seven days after being born. The virgin queen bee, like her human teenage counterparts, wants to hit the spots

DAVE DOROGHY

in town where she will find boys interested in meeting girls. When she is ready to go on the prowl, she leaves the safety of the colony and flies off for her big fling. Unlike younger teenage girls, who tend to huddle in giggly packs, queen bees fly solo.

As with all successful dates, a bit of prep work is required for the queen to land those perfect partners. First impressions really do matter in the insect world. Before she goes on her lengthy mating flight, the young queen will have taken a few shorter flights to help her get in shape and strengthen her wings. Heading into a marathon airborne sex orgy, the last thing you want is weak wings that give out and ground you. Plus the exercise makes her svelte and more attractive. You get the picture: the queen bee is no different from many sexually active men and women who hit the gym to slim down, tone up, and increase their dateability. When the queen is fit enough, she considers how to make herself even more alluring in order to seduce the perfect drones. For colours, she usually sticks with the basics: yellow and brown. Then out comes that naughty, irresistible, seductive perfume pheromone. A sexually active queen bee naturally exudes this pheromone, powerfully broadcasting the fragrance from her mandibular gland. As the queen flies through the drone congregation area, she secretes her unique "come hither" pheromone guaranteed to attract and fire up all the drone bees she encounters, sending them into a sexual frenzy.

Arriving at the drone party after a long, rigorous flight from her home hive, the debutante plays hard to get until the perfect strong, handsome drone shows up. She is looking for her very own stud, a boy bee with a certain buzz and flair and a big, muscular thorax—an enticing, skilful tiny dancer to push all the right buttons at 100 feet above sea level. When the perfect drone shows up, she flitters and flutters around for a very short time, erotically teasing him by winking one of her five eyes at him and sensually flicking her long, wet proboscis. The mood has to be just right with the sun high in the sky, romantic wisps of puffy white clouds far off in the distance, and the twittering sounds of robins and chickadees. At that perfect time the fastest drone catches her; he flies slightly above and behind her, grabbing her abdomen with his legs.

That is the extent of bee foreplay, however. She quickly seduces the drone she desires, enticing him to insert his barbed endophallus penis into her bee "vagina" at the end of her abdomen. This, of course, is exactly what the horny drone has wanted to do, has been programmed to do, and was meant to do since the day he was born and crawled out of his tiny six-sided wax cell. As the 12 insect legs intertwine in mid-air, the drone, caught up in the moment, is so ecstatic he loses his mind.

Overheated with lustful and unabashed insect desire pulsating through his tiny pelvis and into his detachable penis, the love-starved drone can't control his deep-seated sexual instincts. All this love bug can think of is kinky mid-air intercourse with his beautiful mistress cloaked in risqué, soft, sweet fuzz and clouded in pheromone. Bees don't speak words, of course, so one can only imagine what a drone would say during this intimate moment. Perhaps he'd utter her name, "Queenie," or buzz a sweet nothing in her ear as the two become one, or maybe he would just groan, "Oh, honey." Whatever he says, they would be his last words. The desire the drone gives into when he penetrates the queen is so strong, it overrides the fact that this copulation will immediately kill him.

Caught up in high-altitude sexual acrobatics, the drone breathes in more and more of the virgin queen's sweet pheromones. He becomes so aroused he can't stop thrusting himself into her abdomen, finally climaxing so hard it makes an audible popping noise. You can actually hear the sound of a fatal bee orgasm! As the drone literally explodes in rapture, he falls to the ground to die. He dies because his endophallus (internal penis), which he has turned inside out to mate, is barbed for easy, rapid attachment, and as such it remains lodged in the queen's abdomen, so the lower half of his guts get ripped off along with his tiny member—hence the pop.

Sadly he dies knowing that although he may have been the first drone to mate with the queen that day, he won't be the last. She will soon forget him and go on to have many other rapid-fire airborne lovers. The only reason she exists is to advance the ongoing circle of life in the hive and collect as much sperm that day as she can. Love 'em and leave 'em.

But she can't forget him that easily, as the penis he left in her has a smelly mucus coating known as a "mating sign." That sticky aftermath is now all over her rear end. Other drones lured in to mate with the queen later that day will attempt to remove the first bee's mating sign and replace it with their own. Bee sex is raw, messy, and not pretty to look at. As the queen continues putting notches on her bedpost, she gets covered with more smelly mating slime. The indiscreet dalliances continue over and over again, up to 20 times, until she is a slimeball of hormonal secretions. These smells, in addition to her own pheromones, only add to her desirability and allure. During this wild coming-out party, where smells and pheromones mean so much, she will continue to repeat the mid-air sex act with different drones all day before her deep-seated lust is satisfied.

The annual sex blowout I just described is hardly ever seen by prudish and reserved earthbound mortals. It is unusual to see thousands of drones congregating and hovering while awaiting a queen. It is extremely rare to see a queen and a drone actually mating, or to hear the popping climax sound. I have never seen or heard it, and I don't know anyone from our bee club who has, but I saw it once in a documentary film clip. It was in a trailer for a movie called *More Than Honey*. I must warn you that this particular clip is raw, unadulterated wild bee sex. It shows a queen bee embracing a drone in mid-air. Then, close up and in high-definition colour, *bang*! Pop goes the penis. It's really fun, albeit a bit disturbing. The 10 seconds of extremely rare footage were caught by another kind of drone, the mechanical type attached to a camera. You gotta see it. If you have nothing to do tonight, google "queen bee's wedding flight." It's total bug porn! But just watch it once. Like any porn, bug porn can be unhealthy and addictive.

This goes on annually at drone congregation areas all over the world. The female sex bomb flies through the air, systematically sending lusty drones down to their graves while she fills up with their precious life-giving bug sperm. Picture this: as the satiated queen finally heads back to the hive, up to 20 drones will lie dead on the ground, each drone with a perverted smile frozen on his face and missing his puny penis. And those bees were the lucky ones. Every single drone at the congregation area will die anyway in the next few days. At least the lifeless ones on the ground were able to mate, have a second or two of pleasure, and die with purpose.

As for the drones that didn't get laid, well, it's a cruel numbers game designed by Mother Nature—a lot like my odds at the discos in the '70s. Only about 1 in 1,000 drones is successful in hooking up on the elevated dance floor. The vast majority of drones that don't die mating the queen will later return to the hive, where they will be about as welcome as drunken sailors. If there was a shortage of nectar, they will not be allowed back into the hive and will starve to death. No big deal to the female bees, though—the hive will produce new drones the following spring and the cycle will continue.

Here is where the narrative becomes scientific. Although the queen harvested over 6 million spermatozoa from all those rowdy drones, she doesn't conceive on her mating flight, despite the high volume of penetration because she was not carrying millions of eggs with her at the time. She produces eggs in a separate place in her abdomen on an as-needed basis back at the ranch. Keep in mind this is her one big day out, and she needs to fill up with enough sperm to last her lifetime. She will never go out and mate again. She stores all that spunk so she can access the life-giving sperm upon returning to the hive, where she will produce eggs as per the needs of the hive in order to give birth, and give birth, and give birth, and give birth.

This is all part of the bees' circle of life to ensure healthy breeding and to allow the hive to thrive. The queen must continually replenish just the right mix of male drones and female worker bees, especially because they live for such short periods of time.

In the 1960s, ultrasounds became available in North America to help parents determine the sex of their children, among other things. Once again, bees have beaten us to the punch. There is no guesswork in the hive. The queen determines the sex of every single future bee while she lays her eggs. To help her decide, she draws on the thousands of bees in the hive to advise her. Collectively they know the exact ratio of males to females from all the work they have been doing feeding, cleaning, and capping each cell. The proper mix of drones to females changes from year to year, depending on a variety of conditions, including the weather, the season, and the hive's ever-changing flora surroundings. While the queen lays eggs, the workers dance and vibrate around her, communicating and consulting with her on whether to lay a boy egg or a girl egg.

As the queen positions her bum in a cell to release an egg, she has two choices. She can drop an unfertilized egg that will become a male drone bee. Or she can add that sperm inside her body to a microscopic egg and create a fertilized egg that will become a female worker bee. It's nothing short of amazing. She actually controls the sex of each egg she lays, depending on the size of the cell the egg goes into. She measures each cell with her tiny front legs before dropping a tiny egg into it. The big cells are designed for male drones; the smaller cell cribs have been built for the female worker babies—talk about planned parenthood.

As beekeepers, we don't need to wait the 21 days for a female worker bee to hatch or 24 days for a male drone bee to hatch to learn the sex of each bee. We can tell by pulling out the frames of comb and simply looking at the brood pattern of the six-sided cells. Eggs, or "brood," as we call it, are laid in distinct patterns, textures, and locations, depending on the sex of the inhabitant of the egg. The queen will lay drone brood on the bottom of each wooden frame, whereas she lays the fertilized female eggs in the centre portion of the hive. The brood pattern for healthy female eggs is usually a big circle the size of a volleyball in the middle of the frame. Another clue is that the drone brood is always capped with a bumpy, crusty cap, elevated about a quarter of an inch above the cell. This is partly because the drone bees are slightly larger than the female workers and need a bit more room in the nursery. The rest of the female brood is capped with a totally flat surface, resulting in a smaller, cozier crib. So when you see a beekeeper lifting a frame of bees out of a box and studying the frame carefully, one of the things he or she is looking for is brood patterns.

Last but not least, there is the "wild-card egg." We know how male and female bees are made, but where does the queen—the super-female bee and source of all bee life—come from? Don't say Buckingham Palace. The bees in the hive collectively know when their queen is running out of steam, and it's time to make a new queen. Like Queen Elizabeth, queens in a hive will often outlive their subjects. As noted previously, a queen bee lives for three or four years, compared with worker bees, which live for only six weeks in the summer and up to six months in the winter. Eventually, even the queen bee will lay her last egg and die, and so, as with all things in a hive, a succession plan is already well in place. The worker bees sense that a new queen will soon be needed, and so they get busy building a special wax cell for a royal baby bee. Like a luxury condo, it's huge. It's called a queen cell or supersedure cell. They are about the size of your pinky finger and hang down from the face of the cone or the bottom edge, looking totally different from the standard six-sided cells.

Then the worker bees start feeding the larva in the queen cell a special diet that actually changes the developing bee fetus's molecular structure. They feed it something called royal jelly. No, royal jelly is not what Elizabeth and Philip slather on their toast every morning. It's a special secretion that oozes out of the tiny glands of nurse bees. The worker bees actually feed microscopic amounts of royal jelly to all of the larvae in the colony, regardless of their sex or caste. When the worker bees decide it is time to crown a new queen, they feed copious amounts of royal jelly to the little white larva curled up in the fetal position in the queen cell. This supercharged diet triggers a different type of development: the tiny cells morph into a larger queen with the fully developed ovaries needed to lay eggs.

There is one more important step in the birth and emergence of a new queen. Usually the worker bees, sensing a new queen is required, will build several queen cells and feed them all the supercharged royal jelly. This is their way of increasing the odds of producing at least one healthy, fertile queen. After a few weeks, six or seven queens may be produced in a single hive. But make no mistake about it: all hives have only one queen to rule in the end. The newborn queen bee, if she is first to hatch out of her cell, will go over to the other queen cells and kill the developing queens inside. If two or more queen bees hatch at the same time, they will ruthlessly battle it out to determine which one will take the throne. In a cutthroat, winner-take-all fight to the death, only one queen is left standing. It's truly survival of the fittest. After the fighting comes the loving: in 7 to 10 days, the new Queenie goes out for her flight of delight. There you have it—the sexy, messy, and semi-macabre tale of bee dating, mating, and birthing. It's all carefully orchestrated by that sometimes sweet, sometimes cruel, ultimate queen of queens: Mother Nature.

Reflecting on the whole bee propagation process, I can't help but contrast it to my younger, exciting disco days and my modest share of mating flings. Thinking of the tragic fate of the drones, however, I am so glad that my penis is not detachable and barbed. I am still here, with honey in my cupboard and admiration for my bees. Every morning I marvel at the miracle of life and positive energy in the hive. It makes me happy. Eat your heart out, Hugh Hefner.

OLD MAN WINTER

Once spring and summer's virility has passed and autumn's colourful glory has faded, the harsh Canadian winter on the Fraser River presents significant problems for both man and bug. Bitter, icy northern cold fronts will not only challenge a colony of bees cooped up in a small wooden box on an exposed deck for five freezing months, but also test the mettle of a balding bachelor beekeeper hunkered down inside an old wooden barge.

When Old Man Winter arrives, no matter how high I turn up the electric thermostat for my old baseboard heaters, no matter how much firewood I feed my trusty claw-footed cast-iron wood stove, and no matter how many thick grey wool sweaters I layer atop one another, I still can't stop shivering in the humid float home when the mercury plummets and the cold winds blow. During my second winter with the bees, the temperature certainly dived well below freezing on three separate occasions. Each cold snap lasted over a week and turned the usually pleasant, cozy houseboat into a Siberian prison.

It is rare for the Fraser River to freeze, but when it does it is very pretty, especially when a thin layer of fresh snow sparkles on the smooth, flawless ice. The frozen river is as beautiful as it is impractical. A hose, not a lot different from the garden hose you use to sprinkle your lawn, supplies my float home with water. The hose runs from the dock behind me into the river and then resurfaces near the front of the hull, where it is plumbed in above the waterline. When the river freezes, the hose freezes. When the hose freezes, I have no water. All living creatures, including bachelors and bees, need water to survive.

So how do the bees hydrate and get their daily water when the great Canadian outdoors is the same temperature as the inside of your deep freezer? It has to do with the water content of the honey inside the hive, because honey is actually a supersaturated sugar solution. Mother Nature is amazing and always prepared for the changing seasons. One of the reasons the bees reduce the honey's water content by flapping their wings to aerate it in the summer is to prepare the honey not to freeze in the winter. The coveted 18.6 percent water content of honey is a level where there are more dissolved solids than are usually found in any liquid. Just think— honey is delicious, it's nutritious, *and* it's antifreeze.

In winter when the water in my pipes freezes, I must go to the grocery store for bottled water to survive. As usual, when it come to survival, my bees are smarter and several steps ahead of me. Since they figured out how to properly prepare their honey for sub-zero temperatures millions of years ago, they can just stay right at home. However, I do need to take a little credit for their winter wellness. I carry out key winterizing tasks to help my girls "make it through December," to quote Merle Haggard in one of his greatest country tunes.

Just like me, bees need a way to stay warm through the winter. As you know by now, bees live in harmony all year without much friction—that is, until winter, when they need to create friction by shivering, which, in turn, creates heat. Even though bees have their own heat-producing strategy, beekeepers must do three things to assist them in staying warm.

First, a beekeeper must usually reduce the overall size of the hive. My hive consists of three wooden boxes containing bees and honeycomb. In spring, summer, and early fall, the population of a three-storey high-rise bee apartment has somewhere between 40,000 and 50,000 bees, and in winter, Mother Nature naturally reduces the number of inhabitants to a mere 20,000. At this point, my hive was tottering around at a mere 10,000 bees. Heating is all about efficiency. With fewer bees in my hive, there was no sense in heating all that extra space. So, first, I pulled off the top box full of bees and laid it down on the deck. Then, ever so carefully, I lifted out each frame packed with bees from the box on the deck and shook the bees into the lower box, thus combining three hive boxes into two overcrowded boxes. Overcrowding a beehive is a good thing in the winter because it means warmth, and warmth means survival. I suppose if I invited 30 or 40 of my friends over to my float home on a cold winter's night, and we all formed a tight cluster in the living room, shivered against one another, maybe to some Bee Gees music, it would heat things up too.

The second task for winter preparation is something that any clerk at the local Home Depot would recommend: insulation. I insulated my hive with good old-fashioned one-inch-thick pink Styrofoam. I recommend the Dow Cladmate brand, and I did some extensive research before investing the $23.95 to buy a sheet of it. After plugging the ever-popular, trending term "beehive insulation" into Google, I found out that Dow Cladmate Extruded Polystyrene Insulation helps manage energy loss and moisture, it's lightweight, the sheets are easy to cut, and it has an R-value of 5. I had no idea what an R-value of 5 actually meant, but I assumed it was better than R-4 and not as good as R-6. Most people who buy Dow Cladmate at Home Depot buy hundreds of square feet of it; however, one sheet was fine for my small beehive winter renovation project. When I got home, I cut the sheet into five easy pieces with an X-acto blade, then I used old inner tubes from my bike tires as makeshift bungee cords to wrap around and securely fasten the insulation to the outside of the hive. In less than 10 minutes my bees were as snug as bugs in a rug. It was a heck of a lot easier than properly insulating the float home, which is something on my to-do list that I've never quite gotten around to doing. I take the lazy approach of turning up the thermostat, throwing another log on the fire, and donning a big sweater.

The third crucial task to help the girls stay warm is to reduce the humidity and moisture inside the hive. When it gets cold, the bees' biggest enemy is moisture, and my bees had it particularly rough living only three feet above the water. As anyone who has ever spent a winter in Palm Springs or Phoenix knows, when the air is drier, somehow it doesn't feel as cold. There is nothing worse than damp, damp cold. I replaced the now-empty box with a narrower, four-inch wooden box to top off the remaining two boxes crammed with all of my bees. But instead of filling the new, thinner box back up with honeycomb frames and bees, I filled it with wood shavings to insulate the hive, trap moisture, and provide a bit of ventilation. To keep the bees separated from the wood shavings, I stapled a layer of burlap from an old potato sack to the bottom of the top box.

Here's the rub—literally. Inside the hive the bees were now living nice and cozily in smaller quarters, so cozily they would consolidate into a clustered, buzzing ball. That ball of oscillating bees becomes the equivalent of an electric baseboard heater and wood stove combined, and come to think of it, the hive probably ends up warmer than my float home. The warm air created by the bees, along with ambient moisture, of course, rises. When the heated air reaches the upper box with the shavings, the wood chips absorb the excess moisture, and thus the bottom hive boxes stay nice and dry. Theoretically.

Once I created that moisture "sucker-upper," I also adjusted the opening of the hive's bottom board and main doors—which I had reduced in the summer to keep out evil intruders—to provide further ventilation. Think about being in a small club on a crowded dance floor and how nice it is when the owner finally props open the back door to let in some fresh air. Finally, I tilted the entire hive slightly forward by putting a couple of shingles at the back as shims so that any condensation that dripped to the bottom could flow out the tiny door to the front. Moisture in a hive (or a float home) over the winter is not good. Aside from increasing coldness, moisture also promotes slime, diseases, and mould. A dry hive is a happy hive.

When I had completed these relatively simple hive-winterizing tasks, I noticed that as the days grew shorter and we approached the dead of winter, both the bees and I went out less, and this was particularly true during the worst cold snaps.

Please don't get me wrong in this next paragraph. It's not like I turned into a bee that bone-chilling season. This is no Franz Kafka *Metamorphosis* tale, yet I couldn't help but observe that my winter habits mimicked the behaviour of my bees. I stored up food as the roads in Vancouver became snow covered and treacherous. When the first cold snap hit and the weather forecast called for two days of heavy snowfall, I buzzed over to Costco in my vw van and foraged through the colourful aisles. I packed my cart and the van like six-sided bee cells with supplies: family-sized multi-packs of macaroni and cheese, olive oil, canned tuna, eggs, bananas, jumbo boxes of crackers and cereal, bottled water, almond milk, and some veggie burger patties. After this foraging excursion, I didn't have to go out at all for food because I had my winter supply stored away in dry, rectangular (as opposed to hexagonal) cupboards.

The next part of my cold-weather diary is a bit embarrassing, as it deals with going to the bathroom. Flight activity around the hive in winter grinds to a halt like a major airport during a blizzard. The only reason the bees venture outside is to poop. The correct terminology for these trips is "cleansing flights." Bees don't pee. They store as much water as their small bodies can hold and release a tiny bit of liquid waste in the form of uric acid. Uric acid contains hardly any water. When it comes to doing "number two," the girls work hard to keep the inside of the hive clean. As such, each worker bee will hold in her feces until she is well away from the hive. So the main reason for the bees to leave the hive in the winter is simply to fly out and use the great outdoors as their bathroom.

I was in the same boat as my bees during the most brutal cold snap. My pipes were frozen, the toilet wouldn't flush, and I was forced to brave the elements and go outside onto the deck to pee into the river. I know what you are thinking now. Is it bad to pee into a tidal river, and what about his hygiene with no water? I worried too.

According to a cute little video I found online, it is not environmentally egregious to pee in salt water, though the same does not hold true for freshwater sources. Urine contains 95 percent water and the waste by-products urea, sodium, chloride ions, and potassium. Our urea is a bit different in compound and form from the bees' more solid uric acid. The ocean contains the same compounds found in human urea. Also, nitrogen, another pee by-product, combines with sea water to create ammonium, which ocean plants use for food. To top it off, the amount of urea is but a tiny drop, while the great big sea, to where my river travels, measures liquid volume in zillions of litres.

Now, as for my cleanliness during this waterless, cold-weather hibernation, fortunately there is a nice warm community centre and swimming pool (which you are also not supposed to pee in) down the street where I showered every morning and used the toilet. My frozen pipes created other cleanliness problems though. With no water to wash dishes, I simply let the plates, glasses, and pans pile up and then placed them outside on the deck every couple of days. Just like the bees, I did my best to keep the inside of the float home dry and clean by expelling waste products to the exterior.

Occasionally over that winter I'd put on my thick, bright orange down ski jacket, throw the bee suit over the puffy coat, and gingerly navigate the slippery deck to check on the bees. When I did this, I quickly closed the back door behind me so as not to let any more cold, wet river air inside and employed the same technique when I lifted the lid on top of the hive. I was keenly aware of every particle of precious hot air that escaped when I opened the hive lid. My occasional checks verified the girls were still alive. Always a relief. As my own Costco food supplies dwindled, I worried whether my bees had enough honey to get by. When I had reduced the hive that fall, along with feeding the bees the required sugar water to fatten them up, I carefully lifted each frame, estimating its weight. Based on the reassuring heft of the frames, I believed there was enough honey in the hive to make it until spring. A good rule of thumb is to leave a hive with about 40 pounds of honey over the winter. I didn't take an ounce of honey out that winter; I left them every drop. I didn't have one jar of honey to give away that season, or even one for myself. The sacrifices I make for my girls!

Still, I wanted to be sure they had enough to eat. Since sugar water would just freeze in such cold, Jeannie prepared a weird doughy mixture of icing sugar and honey that wouldn't freeze. I placed globs of it inside the hive every few weeks during the worst cold, and they gobbled it up. At winter's end, I also fed them some pollen. A beekeeper at the club suggested I purchase something called a "pollen patty" from the bee supply store. I figured I had my Costco veggie burger patties, so the bees deserved some patties of their own. As the name suggested, they were small clumps of pollen moulded into oily, flat circles the consistency of peanut butter. I am not sure where the pollen came from. It was probably stolen from some other poor hive in the summer and then fed back to my affluent bees in the winter. The beekeeper I gave six bucks to for the pollen patty instructed me to throw a dollop the size of a matchbox on top of the inside of the hive for the bees to munch on near the end of winter. I did, but when I checked again, they hadn't touched it. I, however, finished my entire family-sized box of veggie patties all on my own by halfway through February.

Spring is always glorious—snow melts, rivers unfreeze, and flowers bloom. Brave little yellow crocuses, an array of snowdrops, and fragrant hyacinths pop out of the soil, calling out to the bees: "The pantry is open! Come and get it!" Blossoming flowers are so damned tempting that they seductively lure the hungry bees back outside. I, too, was lured outside. With my houseboat pipes thawed and water flowing again, I no longer visited the recreation centre to use the washroom. I went to Costco less and resumed biking and visiting the local farmers' market. One warm, sunny day in late March, I headed back to my float home after a pleasant forage at the vegetable market and a not-so-pleasant trip to the recycling centre to unload the plastic water bottles that had provided my hydration during the deep-freeze. I couldn't wait to get back to my hive to see what the girls were up to; maybe they were spring cleaning and recycling too.

When I arrived home, I paused for a few minutes on the boat ramp as I spotted about 40 or 50 happy bees buzzing outside of the hive's entrance. Just as I had been out picking up fresh vegetables from the grocer, they were out flying around to gather fresh nectar and pollen. Together we had made it through the harsh winter—bee and man, hive and boat. I imagined that April, May, June, July, and August would be a cinch. When Old Man Winter limps away scowling, and spring and summer waltz back in, accompanied by cheery colours and birdsong, we humans are often filled with a renewed sense of optimism and purpose. It's as if Old Man Winter carries away our worries and troubles in a sack slung over his back, and we get a brand new beginning. I had a couple of years of beekeeping knowledge and many hard lessons learned under my belt. I felt that this would be the year I would vanquish the foes of my hive, and the girls and I could revive and reinvigorate.

ALL MY BEES ARE DEAD

Bee suits don't come in black. Beekeepers wear white because bees can better perceive dark colours and are more apt to feel wary of approaching shadows such as skunks, raccoons, and bears. Garbed all in white, humans can safely approach a hive with the bees becoming less defensive and agitated. Until now, white had been the standard and appropriate colour for my beekeeping attire.

Despite my peppy outlook, it had been another challenging beekeeping year for this bachelor on a barge. With increased diligence, I had applied everything I knew to keep my hive afloat. But the wasps returned with a vengeance, various moulds and mysterious infestations gained stronger footholds, and the queen continued to let her crown slip with insufficient brood production. By the fall of my third year as a novice beekeeper, I'd still not harvested another ounce of honey. I skulked about the neighbourhood, hoping no one would mention the words *bees* or *honey*. I sat in the back row at the beekeeping club meetings, if I even went. I kept mum as Jeannie's hives thrived and grew. As the days grew shorter and the temperatures dropped, the girls and I, ragged and weary, hunkered down for our third winter and hoped for the best.

The following March on a grey, cold winter afternoon, I ventured out in the wind and rain to check on my hive. Suddenly, I wished that bee suits came in black, because my usually lighthearted mood quickly turned to sorrow and regret. My hive was stone cold silent. On that blustery day, I discovered that all of the bees in my hive were dead. Every single one. Normally upon lifting the lid, I paused to simply enjoy the teeming, splendorous miracle of bees buzzing in their happy, cooperative colony. It filled me up. But on this particular grey, overcast afternoon, I lifted the lid of a coffin.

This discovery hit me like a sledgehammer to the abdomen. A three-inch pile of yellow-and-brown corpses lay expired on the hive's bottom board. The reduced hive of about 10,000 bees hadn't made it through the winter. Where once there was abundant sunny energy and springtime joy in anticipation of whimsical foraging flights, now there was a mound of twisted, brittle, lifeless bodies. Order, organization, and purpose had been supplanted by thousands of carcasses waiting for wasps and ants to come and eat them. In the cells where once there was honey, there was mould and the stench of death. I spent three or four minutes just staring at the motionless bees that lay at the bottom of my hive. Then I went inside to call Jeannie.

Modern apiarists face significant beekeeping challenges their predecessors did not. There is a brand new term that didn't even exist 50 years ago: colony collapse disorder. Today, nearly half of all beekeepers' hives are dying every year. My friend Axel is the bee inspector for eastern British Columbia. He is semi-retired, about six years older than me, and works for the provincial government inspecting hives, giving advice, and generally supporting beekeepers in the vastly underpopulated, beautiful farming and mining region of our province. He's the most knowledgeable bee person I know. I heard him give a talk at our club once, and he ended it by explaining how easy beekeeping was when he was a young boy. He said that he and his dad would place five or six hives on the flat roof of their house every spring, do absolutely nothing for six months, and then collect copious amounts of honey in the fall. They would leave the hives outside for the entire winter, again doing absolutely nothing, and all the hives would make it through with no problems. They had strong, independent, hardy bees of the old-fashioned variety that didn't need beekeeper intervention to survive. Of course, 50 years ago the world was a very different place in terms of climate, human population, and bee habitat.

Today a number of bee species all over the world, including the honeybee, are dying. Before you panic, there are over 20,000 varieties of bees on the planet (that we know of thus far). But before you relax and say, "What's a few honeybees then?" consider that the bee species that are struggling with a number of known and unknown environmental and human stressors are in many ways a barometer for the well-being of all pollinators and our ecosystems in general. Their challenges are a distress call to which it might just behoove us to pay attention.

This feels like a good time to share a quote that has been much used and perhaps mistakenly attributed to Albert Einstein. He supposedly once said something to the effect of "If the bee disappeared off the surface of the globe, then man would have only four years of life left. No more bees, no more pollination, no more plants, no more animals, no more man." Again, before you panic, keep in mind the wide variety of pollinators, including butterflies and moths, as well as our dear old friend the wind. Yet, according to the Food and Agriculture Organization of the United Nations, 90 percent of the world's food comes from only 100 crops, and bees are needed for pollination of 71 out of those 100 crops. This is definitely food for thought.

I will now quote myself so that I can claim for all eternity that I am quoted in a book following a quote that may be by Einstein. Dave Doroghy once said, while shivering on the Fraser River one late winter's day, in shock over the demise of his clever, honey-producing girls, "If the bees disappear off the surface of my houseboat deck, then I have only myself to blame. For I am probably one of the worst beekeepers on earth."

After discovering my hive was dead, I plunged directly into the guilt phase of the grieving process. Jeannie tried to console me on the phone. She had recently harvested over 300 pounds of honey from her four hives, so she said she would give me a few jars. She also helpfully pointed out that although I had made a lot of mistakes with my beekeeping, at least I was learning. She said she really liked the colourful cartoons I had painted on the outside of my bee boxes. I felt like a five-year-old. After we hung up, I went back outside, hoping that a few bees might still be alive. Now I was clearly in the denial phase.

I scanned each frame with care and realized to some degree what firefighters must experience when they have to search through burned-out buildings for bodies. Each wax cell, once a small room functioning as an incubation unit, a nursery, or food-storage area, was now nothing more than a cold morgue. Dead bodies were strewn across the surface of the frames and embedded inside the chambers.

Parts of the hive looked like the ancient Italian city of Pompeii, which was buried under tons of ash from a volcanic eruption. During an excavation nearly

2,000 years later, archaeologists were able to see the exact position people were in when they died. My stomach went into a knot when I spotted a bee corpse with its head poked into a cell; it had obviously been feeding on whatever dismal stores of honey were left. I saw a three-quarter-formed larva in the fetal position, curled up dead. I saw a poor little bee with its legs stretched out in the position bees take when they are forming wax comb. And I saw an innocent newborn bee . . . I am sorry; I just can't go on. Contrary to my last ray of desperate hope, there was not one survivor. I felt like I was going to throw up.

Returning to the warmth of the float home, I kept thinking that all the bees just couldn't be dead. Surely some must still be alive buzzing around somewhere, perhaps just late getting back to the hive? Maybe they would return tomorrow? I was like a cat owner whose pet has disappeared for three weeks and is still convinced Tabby will wander home, when in reality Tabby has been eaten. Later that night I realized that I was stuck smack in the denial phase of the grieving process. As Martin Luther King Jr. once said, "We must accept finite disappointment, but never lose infinite hope." I even set the alarm on my iPhone in order to wake up at sunrise the next morning in hopes of finding that the bees had returned overnight.

Next came the anger phase—you can bet it hit me hard. I was about $2,000 into this whole beekeeping venture. Other than the first year of good luck after my sister had dropped off the hive, that fortunate first year when I hit the honey jackpot and we won the prestigious award, I had not harvested one drop of honey. And now, after all of my investments in books, bee school, conventions, and clubs, as well as numerous pieces of equipment, two new queens, and frames of brood, now, after a number of stings and endless embarrassments, all of my bees were dead. You bet I was mad. As angry as a hive of bees toppled over in the back of an old pickup truck on the way to the outyard, as angry as a guard bee warding off an invasion of sinister hornets. Not to mention that the two grand could have bought me 400 small jars of honey, enough honey for the rest of my life. All I had to show for my beekeeping was a bunch of empty, ironically colourful bee boxes, a beekeeping suit that leaked, and a wounded ego and heart.

Next came the depression phase, and thank goodness this dark abyss lasted for only one morning. The different emotional grief phases came on astoundingly quickly. After I got up the next day and was forced to accept that no bees had returned to the hive, I had to dispose of the thousands of dead bee carcasses. What a macabre task. I had affectionately called these little corpses "my girls." Usually when you lift boxes of bees you do it with a degree of fragility, with slow, careful purpose. The empty boxes on my float-home deck no longer merited that type of painstaking treatment. I quickly disassembled the three-storey death house, anxious to get the extremely unpleasant task over with as soon as possible.

What followed had all the dignity and decorum of taking out the garbage. There is no way to dispose of 10,000 bees in a respectful, solemn, and meaningful way. Yes, it was sad they were all dead. After all, they were individual bugs. Each bee had been a unique living being in her own right. But what was I supposed to do? Pick them up individually and say my goodbyes to every deceased bee? Throw the bees into the river one by one? Aside from the impractical fact that such a ceremonial approach would take all day, I just wasn't into handling them. I did what I suspect most people would do in my situation—I grabbed the bottom board with the three-inch-thick mass of dead bees, held it out over the Fraser River, and turned it upside down. I recited to myself, "Ashes to ashes, dust to dust, and bees to the riverbed." As I watched them all float on the outgoing current down the river, as if they were 10,000 Egyptian queen mummies on their funerary boats heading to a floral-laden bee afterlife, I was glad that none of the resident swans were around, as they probably would have eaten them. Thankfully, the dead bees peacefully floated away down the muddy river and out to the ocean—out of sight, but not out of mind. Needless to say, I was quite unmoored for the rest of the morning.

As for the guilt part of the grieving process, I was most certainly guilty. Even with all the time and money I had invested, I was still an incompetent beekeeper.

The final stage of the grieving process, acceptance, took me the longest to achieve.

I had gained quite a reputation on the river as a beekeeper; I had even given honey to many of the other float-home owners. I'd often run into them in the small suburban town centre near the river. Errands in town tended to be quite social, exchanging pleasantries and small talk. Only problem was that small talk almost always led to questions about my bees. Imagine how you would feel if you just lost 10,000 of your loved ones (even if they were a cross between livestock and pets), and the next day some well-meaning neighbour confronts you at the supermarket with a big smile, barking out across the aisle, "Hey, Dave, how are your bees doing?" Like the rigor mortis that had settled into my poor dead bees' bodies, I affixed a frozen, deadpan smile on my face. "They are doing just fine," I lied. I couldn't accept that my bees were all dead. I couldn't bring myself to say it out loud in such an informal setting. Sometimes people I ran into would ask, "Hey, as soon as you get more honey, can we buy some?" I would reply, with a regretful shake of my head, "I heard the nectar flow won't be great this spring, so I don't know when I'll have some more to give you. But I'll keep you posted." I even entertained the dismal idea of going down one of the aisles of the supermarket to purchase some of that cheap corn syrup–infused honey from China to pour into my jars and give out to all my neighbours to make them shut up and leave me alone.

Or I could move. After all, I lived on a floating home, and they are pretty mobile. One call to Brent, the towboat guy, and I could be headed down the river following the trail of my dead bees to another dock on another bend of the river.

It was pathetic. I was a living lie, pretending to be an award-winning, government-certified apiarist. Pretending to be Mr. Natural: your friendly neighbourhood beekeeper, your source for fresh honey, one of the top students in bee school. Oh, come on, who was I kidding? The award-winning honey credit belonged to my sister and brother-in-law, and I cheated my way through the apiarist exam. Because of my distracted nature, the shortcuts I took, my sloppy ways, and my inattention to detail, I had killed all of my bees. It was my fault. In the three years since Miriam and Len had dropped off that hive, I went from hero to zero. Of all the grief stages, guilt grabbed me by the collar, shook me, and wouldn't let go.

As a result of my shame I didn't go out shopping for a few weeks, instead living off of my usual Costco haul of jumbo, oversized stashes of everything from cereal to macaroni and cheese. I was fine—I would survive—but the bees weren't coming back, and there was nothing I could do. Beekeepers have a term for what happened to my hive: they call it a "dead out." I have a term for it: I call it "loser on a barge."

I beat myself up for a while longer. Like a father rearing his children, I had wanted the girls to grow up to be healthy and happy. I wanted what was best for my bees. Where did I go wrong?

Then one day I stopped feeling sorry for myself, taking uneasy solace in the statistic that nearly 50 percent of hives kept by amateur and even professional beekeepers are dying each year. What happened to my hive was not unusual. I began to back off on the self-imposed pummelling; instead, I read further into what may be killing our honeybees. Due to the gravity of what I discovered in my research, I decided to take *some* of the blame for my individual hive's demise, but in the larger scheme of things, the blame for the declining state of honeybees in the world needs to be spread around. It made me feel a bit better, yet also more disturbed, knowing there are larger forces at work.

Colony collapse disorder (CCD), a mysterious phenomenon, has played a major role in widespread hive decline. While CCD is not exactly what happened to my hive—it was a "dead out" displaying actual dead bees—it is likely that some of the factors at work in CCD preyed upon my hive as well. Colony collapse disorder first appeared around 2006 and has continued to wreak havoc for over a decade. In a sudden bizarre occurrence, bees began to disappear from commercial hives—no dead bees left behind and no evidence of mites, pathogens, or predators. Even weirder, the queens and brood were left behind. And by now you know the protective and diligent manner in which the workers normally serve their queen. Uncommonly all alone, the queen would wear herself out laying more brood to try to keep the colony alive. But even the amazing egg-producing queen bee could not lay fast enough to replace an entire hive of workers and drones. And, so, the terrifying result: large numbers of dead hives. Sometimes 500 hives could be lost seemingly overnight. Once, over a few months, Bret Adee, America's King of Beekeepers, lost 40,000 hives—about 2 billion bees. And this has happened to him more than once; in 2016, Adee lost about half of his 90,000 hives.

Perhaps I should stop second-guessing every sloppy mite vaporization, every missed wasp, and every incorrectly boiled batch of sugar water. "Dave," I coached myself, "give yourself a break. Part of the reason your bees died was beyond your control."

For a time, scientists looking into CCD were flummoxed, and there is still no hard and fast agreement as to the causes. What we do know is that there is likely a combination of reasons for it. I'll focus on just five or six suspected factors that are easy to understand. Granted, I am just an unsuccessful amateur beekeeper who lost one hive, and I am not an entomologist, but this is my humble understanding of CCD thus far.

Let's start with almonds and the almond milk I buy in bulk at Costco by way of example. I drink large quantities of almond milk because I think it is healthier than cow's milk, and I like the way it tastes. I also rarely leave the float home without a handful of almonds in my hand. Okay, so what do almonds have to do with bees?

California dominates global almond production like the Arab nations dominate world oil production. Over 80 percent of all the almonds produced in the world come from that one state. It's remarkable the way almonds are mass-produced down there in the Golden State, and it is why I can go to Costco and buy a three-pound bag for $30. Our incredible honeybees have everything to do with the success of the California almond industry. Every spring it takes 1.6 million commercial hives to pollinate California's almond crops alone. Let me repeat that: not 1.6 million bees, but 1.6 million *beehives*. And this is just for almonds. In addition, bees are trucked all over North America for agricultural pollination—from

Florida and the Midwest to Washington to pollinate that state's cherries and apples. This is not new. The Egyptians would float their hives down the Nile on barges. But the staggering numbers of bees and hives trucked about is a fairly recent state of affairs. The hives are required for only a short period of time when the trees are in bloom and rife with pollen, and then they are loaded up and off to the next destination. I envision each hive as a sort of ragged, exhausted touring rock 'n' roll band, constantly on the road and only vaguely aware of the next city they must perform in.

So for almond trees in bloom each spring, about 1.6 million hives—estimated to be more than half of the managed hives in all of North America—are trucked in from afar. Most commercial beekeepers across America, and even far away up here in Canada, make more money renting their beehives out for pollination than they do selling honey. Their local honey profits, sadly, have been upstaged by the sugar-infused honey from overseas.

Jeannie and I once cycled through the San Joaquin Valley in California, where weather and soil conditions are ideal for growing almonds, and we saw the almond extravaganza with our own eyes. We usually cycle at about 12 miles per hour, and for an entire hour we witnessed row upon row of identical almond trees as far as the eye could see. Every third row had a four-box beehive sitting neatly at the foot of a half-mile-long line of trees. And what we passed through was puny in comparison with the size of the overall almond industry. In 2015, the California Department of Food and Agriculture estimated that the state had over a million acres of almond farms. Each one of the almond trees on those huge farms is looking for a honeybee to drop by, pick up some tiny bits of pollen dust, and drop them off on another almond tree down the line. Almonds require 100 percent bee pollination. Most other plants can rely on the wind and other plants to help.

But it's really not nature's intention for us to move bees halfway across the country and back again as California pollen couriers, just so frugal shoppers like me can save a few bucks on a bag of almonds. Just like it disturbs our natural rhythms to hop in a car and drive across the United States, it does the same thing to the bees. There are negative stressors, and the bees have to adapt quickly to new weather patterns, new temperatures, and the different position of the sun in the sky. Also, just like travel exposes humans to colds, flu, or allergies, all that travel also exposes bees to unfamiliar pathogens, fungi, and other environmental influences. Above all, the steady diet of only almond tree flower nectar and pollen every day is unnatural. Hey, don't get me wrong, I love almonds more than anyone, but I don't think I could live on nothing but almonds.

So is the answer to saving the bees to stop buying almonds at Costco? Well, sort of, but not really. Without the almond industry, the honeybee industry in its current form could not exist. It's complicated. Colony collapse disorder is a catch-all phrase that points out all the man-made interventions we are creating to screw up the climate and honeybees and life here on the planet. It all boils down to overpopulation. You can cram only so many bees into a hive and only so many people onto this planet.

A second suspected cause of CCD and the downfall of our little pollen-seeking friends also relates to farming. Modern industrial fruit and vegetable agriculture uses a method called monoculture. All that fancy four-syllable word means is the cultivation of one single crop. In the days of yore, small family farmers planted and worked dozens of different crops that would ripen at different times of the year, providing an ongoing living for the family and a continuous smorgasbord for bees. It was not the least bit unusual for beans, corn, carrots, wheat, and potatoes to be planted side by side in perfect harmony. Luckily, many smaller farming operations currently focused on sustainability have revived this practice. Furthermore, before our relatively modern obsession with lawns and lawn mowing, homeowners used to have chickens, goats, flowers, and vegetables, not perfectly groomed grass, in the front yard. Even larger farms used to leave large swaths of tall grass and flowers—a haven for bees—right next to the cultivated crops. But now, in order to keep "pests" at bay, many industrial farms leave no wild corridors at the edges of their fields.

Monoculture is not confined to almonds. On that same biking trip through California, Jeannie and I saw countless other large-scale monoculture crops. This method has emerged from large-scale farming practices aimed at high yields and efficiency, from big businesses concerned with profit margins, and from a human population bursting at the seams needing to be fed. Almost every "perfect" supermarket fruit or vegetable you buy today—and we've grown spoiled in expecting these items any time of year—is grown on a corporate farm that takes advantage of the money-saving economies of harvesting a single crop. Farm-stand fruits and veggies grown locally on small-scale farms that raise a variety of crops with fewer or no pesticides are never "perfect" in terms of colour, shape, and blemishes. I am probably not telling you anything new about how food is mass-produced and farmed today, but stop and think about it from a bee's perspective. To bees, the two farming practices are the equivalent of the difference between a delicious and varied all-you-can-eat organic buffet and an institutional dinner tray with mashed potatoes and creamed chipped beef topped with a dressing of pesticides, again and again and again.

I haven't even mentioned genetically modified food crops, the infamous, potentially evil GMOs, as a source of pollen and nectar for bees. Since GMOs are a complex topic of debate, let's just say that neither monoculture nor GMOs are natural and leave it at that. One thing is certain—when Mother Nature designed the amazing range of bee species millions of years ago, their diet certainly didn't include the kind of food we are producing and feeding them today.

One particular type of sustenance for commercial bees is highly suspected of contributing to incidents of CCD. You guessed it; it's that white processed sweet of sweets—sugar. Beekeepers are trucking commercial bees all over tarnation to slurp on monoculture crops. At the same time, wild foraging habitat are growing scarce, so beekeepers have increasingly turned to using sugar-water solution. Unfortunately, it's become more and more apparent that bees cannot survive on sugar alone. Like humans, they need a balance of carbohydrates and protein. Carbohydrates come from sipping honey in the hive, and they get their protein, ideally, from floral pollen. Due to their hectic schedules and having an increased fare of cane sugar, our honeybees are growing malnourished, which, of course, puts them at increased risk of all the other stressors like diseases, mites, and fungi.

The next possible human interference in the fate of bees is less organic, more high-tech, and fairly controversial. If you are starting to feel guilty because you eat almonds and consume produce from monoculture crops, then this next paragraph comes with a warning: please turn your cellphone off now.

When CCD first appeared, research attention turned to technology. A Swiss study found that signals associated with cellphones confused the highly sensitive nervous systems of the poor bees. Upon exposure to cellphone signals, the bees piped up the way they do when anything disturbs the hive or when they are about to swarm. This original theory linking cellphone signals to CCD has since been called into question by further studies. However, think about this: there are countless frequencies, not just cellphones, beaming through the sky at any given moment in our crowded atmosphere, many of which were not present at all or to the same degree half a century ago: AM radio signals, FM radio signals, shortwave radio signals, radar, microwave signals, and Wi-Fi signals. The list of frequencies and signals goes on and on. Bees communicate through buzzing and frequencies, along with waggle dances, so our human technological noise must be, at the very least, causing one massive bee migraine. I am no scientist, but I can't imagine how this mass of radio frequencies and electromagnetic static is not severely disrupting bee communication.

And it goes deeper than communication. Bees have an innate, finely tuned electrical sense. In relation to one another, bees are positively charged and flowers are negatively charged. This means that when bees pollinate, the electrical polarity helps the pollen stick to bees' hair. The interference we are creating with all of our gadgets and 21st-century device addictions can't be a welcome addition to the bees' well-being.

Unsurprisingly, pesticides are another major suspect in the CCD mystery. Of the wide range of pesticides currently used to eradicate the various pests that love and thrive on monoculture, neonicotinoids have been established as having a strong link to bee decline. Note the word *nicotine* buried in there. A pesticide with this name is not going to turn out well for bees.

The same players that brought you GMOs and monoculture have come up with a new class of insecticides chemically related to nicotine. Only these poisons don't come with a warning for bees written on the sides of flowers, like on the sides of cigarette boxes; there is no surgeon general to warn them. The term neonicotinoids basically means "new nicotine-like insecticides." Like the nicotine we take in when we smoke, the neonicotinoids act on certain receptors in the nerve synapses of invasive insect pests. Neonicotinoids, which are much more toxic to bugs than they are to animals, birds, or people, have also been linked to greater bee susceptibility to the villainous varroa mite. One of the reasons neonicotinoids are so popular is that they are easy to use—their water solubility allows for easy application into the soil, where they are slurped up by plants. In turn, bees hungry for flower nectar drink up the pesticide. Think of it this way: every time bees go in for a sip of nectar, they have to take a deep drag on a cigarette. See, I told you this one was not going to turn out well for the bees. Since birds rely on bees for some of their food, it is easy to see how this little pest-control cocktail might have a negative chain reaction.

Studies have shown that although low-level exposure to neonicotinoids does not directly kill bees, the chemical may impair their ability to forage for nectar, learn and remember where flowers are, and find their way home to their hives. Unlike people, who can choose to kick the cigarette habit, bees can't choose to kick the neonicotinoid habit. They must rely on the decisions that modern farmers and lawmakers—understandably concerned about profits and re-election, respectively— make about spreading more of these pesticides on crops. You can almost hear our bees coughing louder and louder. Fortunately, some countries have passed regulations limiting neonicotinoid use. However, it is not time to rest easy, thinking the problem is solved, because there is often pressure from big agriculture corporations to reduce regulations on pesticides; thus, the way that neonicotinoid use is limited depends on the individual regulations passed. Also, not all countries have limited neonicotinoid use. According to some sources, the theory that neonicotinoids may be a culprit in CCD is even under hot debate. See how complicated it is to try to sort out my poor girls' fate and my hive's dead out?

Many interconnected factors are suspected of contributing to CCD. The final factor is a huge global conundrum: climate change. If temperature fluctuations from a changing climate cause the snowcaps to melt earlier in the year and the flowers to emerge and bloom earlier, it's not clear if the bees will adjust easily. If the flowers are available early but there are no bees around to pollinate them, it doesn't take an Einstein to figure out there is going to be a problem. To make one pound of honey, bees need to visit about 2 million flowers. If the bees miss the first few weeks of blooming flowers because no one told them spring came early, will they have time to visit more flowers later on to make up for it? Will enough flowers bloom later in the season? And what of the effects of potentially increased drought in some areas?

Scientists are quickly trying to piece together and predict the exact effects of climate change on bees. Initial signs point to a detrimental conclusion, which comes as no surprise. Oh, and did I mention that climate change could also pose a few problems for mankind as well? Season length, temperature, and available water make up the essential recipe for growing our food crops. I am sure you've heard the predictions. Climate change will mess with, and is already messing with, this essential recipe. If I were you, I'd stock up on almonds.

In the end, after considering all of these bee life hazards, I felt a bit less distraught about my own ineptitude. With so many complex factors, it is impossible to pinpoint the exact cause of my dead out. Was it my poor beekeeping skills or was it some of the contributors to CCD? It was probably a bit of both. The culprits connected to CCD are weakening hives in general; my hive was just a tiny microcosm of a global epidemic.

After ruminating over the loss of my hive for far too long, I finally came to the acceptance stage of the insect-grieving process. Perhaps it's not that I am a lousy beekeeper after all; maybe I am just a novice who took up beekeeping a bit too late in life. I thought about my resident bee expert, Axel, as a teenager in the mid 1970s. This was before genetically modified foods, climate change, cellphones, monoculture, and neonicotinoids. It seems we may have circled back to Rachel Carson's SOS about DDT in her 1962 book, *Silent Spring*. Here we are again—humans bumbling about and causing problems in Mother Nature's perfect system. I pictured Axel's hardy bees long ago in their hives next to bountiful open spaces full of flowers in British Columbia's West Kootenays, with snow-capped mountains in the background and air so fresh their tiny lungs probably screamed with joy. Axel's bees were independent and needed little human intervention. I fantasized that if I had been a beekeeper then, I'd probably have so much honey, it would make my float home list to one side.

My sadness lifted, if only momentarily. But the blues descended again when I pictured poor hobby apiarists setting out to keep bees 50 years from now. Will they struggle to raise sickly, lethargic bees? Will they only be able to raise bees in giant indoor greenhouses? That was when I powered off my cellphone, shut down my Wi-Fi router, turned off the radio, and went to tend to my non-GMO, pesticide-free flowers on my houseboat deck. I gave each plant a generous drink of fresh water to keep it healthy, so that it could offer nectar to whatever living, vibrant bees happened by on a flight down the Fraser River to their foraging grounds.

SPLITSVILLE

We all face forks in the road. Life is nothing more than a self-directed joyride continuum of decisions large and small. Take that job overseas, go back to university, get married, buy that house, buy that car, bike to Mexico. These are some of the bigger decisions. Return that turquoise sweater, buy a lottery ticket, start a diet, read every book ever written by Ernest Hemingway, make the bed. These are the smaller decisions. The road I'd travelled over the last few years led me to a medium-sized decision: whether or not to continue beekeeping.

Quite frankly, a big part of me wanted to quit beekeeping. I had some long overseas trips planned that I didn't want interrupted by having to come home and feed the bees sugar water so they could survive the winter. I was sick and tired of all the money I was pouring into this money pit of a hobby. I had recently purchased a new vaporizer that cost over $150, and my "old" beekeeping suit had become so full of bee-sized holes that I had to invest $150 in a new and improved one. What did I have to show for my toil and trouble? For the money that had flown from my wallet like a queen bee on her sex-crazed mating flight? I had a ghost town of a hive and no honey.

And one more thing: the novelty of getting stung had worn off. I'd been stung in the face over a dozen times. My ears, nose, throat, and cheeks had ballooned up to disproportionate sizes. In general, not a good look.

As with many decisions, there was the practical analysis: time and money invested versus satisfaction and fulfillment derived. But the analytical part of decision-making so often becomes crowded with emotions: love, attachment, grief . . .

Most of us have experienced the death of a pet—a cat or a dog we loved so much and grew so accustomed to that we are profoundly affected by its passing long after the animal is gone. The ways we process grief after losing a pet are complex and varied, and can sometimes be as intense as losing a family member or friend. Many animal lovers, caring, lifelong pet owners, give this advice: don't get a new pet too quickly. Give yourself six months to a year to get over the loss, and then, after some time has passed, consider a new pet.

I reflected on this after losing my bees. Should I replace them? Maybe Mother Nature was trying to tell me that beekeeping wasn't for me. Maybe she had given me a sign that I should just move on.

But I had all this expensive beekeeping equipment and fancy protective clothing. My close circle of friends and family were composed of beekeepers, and all the folks I ran into at the grocery store now identified me as Dave the Apiarist. Plus I had the official framed provincial beekeeping certificate. Beekeeping had become a part of my identity; it seemed a shame to quit after I had invested three years of my life in learning apiary skills. I wondered if I could recapture the interest and curiosity I had when Miriam and Len first dropped off the new hive on the back deck of my float home.

For days, then weeks, post-bee-mortem, I wrestled with the question of whether this hobby was worth pursuing. I seriously considered that it might be time to let go and move on.

If it was time to let go, then so "bee" it. After all, I had quit or given up on different sports, hobbies, and even animals in the past. It was nothing to be ashamed of; we humans change and grow. Our interests and goals shift. It's perfectly natural. In my 20s, I was totally into coin collecting and playing racquetball. I invested a significant amount of time, money, and energy into these pastimes, enthusiastically embraced them, and then, for some reason, I just quit. I never looked back, and I don't regret letting go of either former passion. But my choice to continue keeping bees was not as simple as whether I was "into it" or not. There were deeper emotional nuances.

Certain past pet traumas were fuelling my bee quandary. Fifty years earlier, my grandmother from New York sent us four turtles in a cardboard box—real turtles, mind you, not the chocolate kind. Manhattan pet stores in the 1960s had adopted a ridiculous trend of painting flowers and other colourful motifs on the shells of small pet turtles in order to create demand. The pet-store owners used indelible oil paint. At the tender age of eight, I learned three useless turtle-shell facts: oil paints contain harmful turtle-killing chemicals, turtle shells are porous and absorb those chemicals into the creature's bloodstream, and a turtle with a cute yellow daisy painted on its back is a turtle condemned to die within three to six months. That was how long our turtles from Granny lived. During that short period, my sister and I cared diligently for our turtles—we read about them, fussed over them, carefully and regularly fed them, and generally bonded with them.

In truth, as far as pets go, the turtles weren't that great. Unlike bees, which fly around like crazy and are fun to watch, the turtles were kind of boring. They moved so slowly we often thought they were dead. Then one day: "Hey, wait a minute. They *are* dead!" Turtles can have only so much poisonous paint flowing through their tiny turtle hearts before they finally croak. Their death in the name of turtle-shell art pretty much closed the chapter on my sister and me raising reptiles as a hobby. I felt like some of my beekeeping mistakes were as bumbling as painting toxic chemicals on a turtle.

Eventually, my ruminations led me to a middle ground. Maybe a break was in order. Heeding the advice of pet lovers on grieving, I figured that taking some time between all the bees that had died and acquiring a brand new hive would provide a valuable opportunity to reflect and heal. After a break, I could start with a clean slate. That's it: a hiatus. But how long of a break should I take? Should I set a time based on the advised six months to a year? Only problem was, that transition formula was for a single pet. I had just lost 10,000 bees; I wouldn't be emotionally ready to get a new hive for at least 5,000 to 10,000 years. That would mean taking up my limited apiarist skills again sometime between the years 7020 and 12020. Heck, even if by some miracle of biology or physics I was still alive, I would have forgotten everything I'd learned by then; after all, I'd forgotten almost everything I'd learned just a week after taking the bee school exam. What difference would it make anyway? I won't be around in 7020, and based on the increased rate of CCD, there is a good chance bees won't be either. And if the bees won't be around, neither will humans . . . and, well, you get the picture.

I decided that if I was going to continue beekeeping, it was best for me to start right away. I needed to quit moping around like a blood-poisoned turtle with a toxic sunflower on its back and roll up my white bee-suit sleeves, take stock of my previous challenges and errors, and refill my empty hive with new buzzing life.

Where to begin though? How could I make sure I didn't repeat my mistakes? I don't ever want to encounter 10,000 brownish-yellow corpses again. Miraculously, I remembered one lecture from bee school focused on certain diseases that remain in wooden hive boxes long after the bees are dead, and this gave me a clear place to begin: my equipment. I recalled that if you don't take action immediately after a dead out, the lingering diseases will go on to kill any future live bees you introduce to those hive boxes.

One disease, American foulbrood (AFB), is so deadly, it requires a blowtorch to scorch the inner walls of the hive to rid the wood of a microscopic spore-forming bacterium called *Paenibacillus larvae*, which can remain active on beekeeping equipment for 70 years. That's right, 70 years! The rod-shaped AFB spores are minuscule and impossible to spot. In a hive infected with AFB, nurse bees feed the poor unsuspecting, newly hatched bee larvae the heavenly brood honey, but it is contaminated with *Paenibacillus*. The babies die within a few days. The bodies of the baby bees decompose into a telltale ropy, smelly (hence the name foulbrood) composition, releasing millions and millions more *Paenibacillus* spores. More babies eat more contaminated honey, the bees out foraging spread the spores, and so on. According to the Ontario Ministry of Agriculture, Food, and Rural Affairs, all it takes to infect a one-hour-old larva is one spore. And it takes only 35 spores to infect a day-old larva.

One way of knowing if you have American foulbrood is by checking to see if the dried, dead larvae have formed hard, brittle, dark scales along the lower walls of each wax cell; these are the remains of what once had promised to become hard-working honeybees. So I pulled out my old frames and inspected the empty cells for signs of foulbrood. I couldn't spot any scales, which was a relief, especially since each scale can contain up to 2.5 billion spores. I was lucky; it was likely I would have accidentally set my float home aflame trying to scorch the bee boxes.

In all probability, my bees, weakened by those dastardly blood-sucking varroa destructor mites, had died of chalkbrood or nosema. I couldn't really remember the difference between those two maladies. It had been almost a year since the bee course. But, luckily, I did remember the instructor mentioning a cool thing called an Iotron electron beam sterilizer that would zap chalkbrood and nosema cells and kill all the microscopic pests that might inhabit your frames, boxes, and beekeeping equipment. I liked the name of the machine; it sounded like something out of *Star Trek*. I figured everything about beekeeping up until now had been so low tech, it would be fun to experience something high tech for a change. If I was going to do this again, I was going to do it right. Beam me up, Scotty!

As luck would have it, Canada's largest Iotron electron beam sterilizer is located in a suburb of Vancouver called Port Coquitlam, an hour's drive from the float home. So one Saturday morning I loaded my contaminated hive into the van and made a beeline to the sanitation station. Arriving in the massive parking lot full of 18-wheelers, I extracted my three-box hive with the 30 dirty frames and, with my rusty beekeeping tool in hand, walked toward one of the loading docks of a huge grey warehouse about the size of three Costcos. I rang the bell, feeling a bit like Dorothy ringing the bell at the entrance to the Emerald City in *The Wizard of Oz*. Shortly afterward, a bald guy in a white lab coat, eerily reminiscent of the Wizard, answered. I fully expected him to say, "Go away!" After seeing my hive boxes, he understood I was legit, smiled, and let me in.

The electron beam accelerator was massive, about as large as a mid-sized commercial airliner. The 15-foot-high, weird-looking machine had a big black rubber conveyor belt running smack dab through the middle of it. I saw huge boxcars of stuff being piled onto the conveyor belt and watched forklifts loaded with pallets of rocks racing around the warehouse. Taking it all in, I realized how small my paltry, quick-fix beehive cleansing job was in the scheme of things here; maybe I should have stayed home and stuck my hive boxes in the microwave.

I was truly interested in just what the hell this machine did and its commercial applications. The guy in the lab coat turned out to be friendly, but he kept on looking at his wristwatch as if he were a bit pressed for time. I didn't take the hint. I broke the ice by asking if he had many beekeepers showing up. He looked askance at my zany, colourfully painted hive boxes and replied, "Beehive sterilization represents less than 1 percent of our business. It's mainly commercial beekeepers that ship thousands of hives at a time down here from all across Western Canada. Some bee clubs will pool orders from their members and ship hundreds of hives once or twice a year." He closed by letting out a little chuckle. "We rarely get someone like you showing up alone with just one hive." I felt so stupid standing there with my single puny hive. Yet it gave me pause that thousands of diseased hives ran through the machine each year; it put my one hive's demise in perspective.

The Wizard must have liked me after all, because he went on to explain that the machine does just what its name implies: it accelerates electrons into a high-speed electron non-radioactive "shower" that sterilizes medical equipment, pharmaceuticals, labware, agricultural feeds, and a host of other industrial items. My ears perked up when he told me a big part of their business is gemstones;

huge mines ship gemstones in massive boxcars from all over the world to go through this very specialized machine. Blue topaz, which is the gemstone of Texas, is actually grey or colourless when it comes out of the ground. After it is blasted with radiation from the electron beam accelerator, it turns an incredibly vibrant turquoise. He told me his plant irradiates hundreds of tons of topaz every year. After a short tour of the plant and wasting more of this poor guy's time with neophyte questions, I was afraid to ask how much he would charge to whisk my slimy, mouldy, and fungus-ridden boxes through his magic carpet ride. "How about I give you a jar of honey and we just call it even?" I joked. He didn't go for it. It was 10 bucks, and I could pick the hive boxes up in the morning.

When I fetched the boxes the next day, they looked exactly the same. Did he really irradiate them or was it the world's biggest scam? Before I left, I stood for a few minutes watching the conveyer belt spin as it cycled through the massive 165-foot-long machine. Since I had shelled out only 10 bucks, I didn't want to overstay my welcome. I packed up my boxes, said goodbye, and left. Exiting the parking lot in my van, I carefully scanned the black asphalt road in hopes of seeing the glint of a turquoise topaz gemstone that had fallen out of the back of one of those huge trucks. A topaz would have made a nice souvenir from my trip to see the Wizard of Oz. Now, with my boxes professionally sterilized and disease-free, I was ready to start beekeeping again. *Carpe diem*!

Port Coquitlam is, as the name suggests, near the water. Taking the scenic route home, I noticed some trees in early bloom along a pretty inlet. No season signals freshness and a sense of optimism like spring. Daydreaming about winter's imminent demise, my thoughts eventually wandered back to beekeeping and my dilemma of restarting my hive. Okay, sanitized bee boxes are great, but what about bees? I was back to square one: I needed some hardy new bees to get started.

I listened to a business report, my attention bouncing from bees to the stock market and back to bees. I like stocks that split, although mine never do. My friend bought lots of shares of Apple stock in 2003, and his stocks have split a few times. Instead I bought BlackBerry shares (fitting for a Canadian beekeeper). With investments like that, I need to save all the money I can and keep my eyes open for deals. When it comes to getting a break on beekeeping expenses, I saved the best for last. Here is the one and only cost-efficient thing about beekeeping: sometimes a healthy hive gives you a two-for-one deal. Remember the chapter on swarming? Well, when you split a hive in two, you are essentially pre-empting a swarm by intervening; the best time to split a flourishing hive is, you guessed it, spring.

Splitting a hive involves simply lifting four or five frames of brood and bees, along with the capped honey and pollen, out of one box and inserting those frames into another empty box. You'll remember that when the bees swarmed, they did it on their own, but they weren't able to take their pantry of stored honey with them. They also weren't able to take along any of the unborn bees slowly forming in the wax cells.

When you intervene and create a split, it is important to take the brood, or eggs, in various stages of development. This way the new hive begins growing immediately and has the staying power to continue growing for the weeks and months ahead. What about the queen though? In the wild, she goes off with the rebels. When you artificially split a hive, just like in the wild, you move the queen with her comforting queen pheromones to the new box to help hold the bees in their new home. The queen will also speed up the growth of the new hive by immediately laying more eggs. But remember, after you move the queen, you will have to add a new queen to the original hive.

Lucky for me, I knew just the girlfriend who had too many large hives and was willing to do a deal with me. Over the previous season, Jeannie's beehives had gone nuts. She went from keeping just three hives on her dad's farm to having her hives grow, morph, split, and swarm into 18 different hives. She just happened to have a super-healthy hive that was ready to split. One nice thing about beekeeping is that it is void of regulations or red tape. No subdivision permits are required and there are no zoning issues to deal with. Mother Nature welcomes these particular subdivisions. So Jeannie and I put on our white suits, reached into the gene pool at her dad's farm, and grabbed what we needed. Then, with bees, eggs, and honey all neatly bundled into a cardboard nuc box, we headed over to the float home. I remarked to Jeannie how her bees had gotten an upgrade from a farm view to a river view.

Part two of the split took place on the float-home deck. As we gently removed each frame of the split from the cardboard nuc box, I was quite proud the new girls were moving into spic and span, freshly irradiated colourful hive boxes—the best 10 bucks I'd ever spent.

For my new bees, it was the equivalent of moving into a spotless apartment where a professional cleaning company had put in 40 hours of elbow grease to scrub those hard-to-get-at areas, like under the sink and behind the couch. Actually, the wooden boxes were even cleaner than that. The ionizer plant sterilized medical equipment used for surgical operations. My hive was as clean as an operating room! Forget about everything that had gone wrong in the past with my half-baked, sloppy apiarist ways. This was the first day of my new beekeeping life.

With a super-clean start, I was finally getting serious about my beekeeping hobby.

As we transferred one of the last frames from the cardboard nuc box into my freshly sterilized hive, lofty philosophical thoughts about the ebb and flow and the circle of life caused me to mentally stray from the task at hand. The day was perfect with a light breeze coming off the river. I paused and looked closely at the single frame I held in my hands. My reading glasses, perched on my nose under the protective veil, were clean for once. With not a cloud in the sky, the light was intense—which doesn't happen often in rainy Vancouver—in a way that illuminated the smaller details that we often overlook. I was totally in the moment. I had a crystal-clear view of a one-inch-square section of the frame. I grew mesmerized by a cluster of six or seven tiny capped brood cells. The frenzied nurse and worker bees rushing to and fro all over, bumping into each other in a hurried Grand Central Station–like manner, had 100 percent of my attention. Time stood still as I observed and tried to understand this contradiction of order and confusion. Then something occurred that under most beekeeping circumstances would be so very easy to miss—something smaller than small and so wonderfully miraculous. A tiny hole, the type of hole that might be created by poking a sewing needle into a wax candle, emerged in one of the cells. Very slowly, a baby bee's minuscule antennae poked out. Then the black vibrating antennae, each the width of a strand of hair, withdrew back through the hole into the safety of the sticky, concealed wax crib. I had stumbled upon the very first seconds of a bee's life. After a minute, she poked her antennae through again, wiggling them for a few seconds to slightly enlarge the hole. Then, again, she rested for a minute. Then she enlarged the hole a bit more with her mandible jaws. She pushed her tiny bee head up against the small wax opening, breaking more wax cap with the pressure. For the next 40 seconds, she bore her way through the hole, making it bigger and bigger. Repeating this instinctive action was, of course, what she was programmed to do. She widened the hole, wiggled back down, paused, wiggled up, carved out more wiggle room, chewed more wax, and paused again. The hole was still too small for her to fit through, so she rested a bit more and then repeated all the steps. After investing four more minutes watching her squirm and push, I was hooked on seeing this birth through to its conclusion.

Unfolding in front of my eyes was the very first step in that complex circle of insect life I had been observing for three years. Obviously, this miracle happens all the time. I had come across the evidence in the past, but I had never witnessed the exact moment that life emerged: antennae poking out for the first time, followed by the tiny bee's arduous destruction of the wax-capped cell as she yearns to join her sisters and begin to contribute to the ongoing important work of pollination and the survival of the colony.

As this baby finally emerged, she looked different from the others. Her exoskeleton and hairs hadn't yet fully developed, and she was slightly smaller and lighter than her sisters. I later learned that her salivary glands are activated when she leaves the enclosed cell and encounters fresh air, which triggers a desire to clean out empty birth cells, including the one she just emerged from. For the next two or three days, she commits herself to housecleaning chores. Other more experienced bees fill the cells with nectar and pollen, after the queen has come by to plop in a fresh egg. The eggs the queen lays, just like human eggs, are preprogrammed with cues, codes, and millions of prompts that result in instinctive behaviour. Mother Nature's mandate to keep this whole world spinning depends in large part on those innate prompts that guide us all. You have them, I have them, and the bees have them.

What this little bee did next totally blew me away. She didn't waste any time in fulfilling her destiny. She crawled out and in less than two seconds began working. What other creature is born and immediately gets busy adding real, tangible value to its society? It took me at least 25 years. Bees are absolutely amazing! This little bee, which had only a couple of months to live on this planet, immediately began contributing to the hive. The lowest job on the bee totem pole, the entry-level job all bees start out at before they get the work experience they need to be promoted, is cleaning. She joined in with zeal and vigour.

Alongside the other bees, she began simply poking her head in and out of the cells, cleaning out bee afterbirth, debris, and old food. It's not like she took a five-minute break to get her bearings and build up her strength. It's not like she needed a tour of that particular frame so she would know her way around. It's not like another bee needed to show her how to clean. She was born to work. My jaw dropped. Then, not to be outdone, I returned to the work at hand: completing the split transfer.

I knew then, and only then, that continuing to keep bees was what I must do. Not for the honey, not for the notoriety, nor to have a hobby to share with my girlfriend and sister. No, I would keep on beekeeping for a reason I didn't even contemplate when Miriam and Len had dropped the first hive off at the float home three years earlier. I would continue with the bees because carefully observing and respecting their world as an accidental apiarist might, in some small way, helps me to figure out how to navigate the complex, much more dysfunctional hive of humans swarming over Planet Earth. Watching bees and taking their collaborative cue might be a way to better understand and connect with our fellow humans. Our challenges are not so different from those faced by bees. We share the same manifest from Mother Nature. We all play a role in how we will deal with the comings and goings of leaders and people, diseases and famine, weather extremes and natural disasters. We are all hard-wired to want the same things: some honey in our pantry and a safe, comfortable home among a hard-working, harmonious colony.

ACKNOWLEDGEMENTS

Thanks to Jeannie, who is excellent at bee- and boyfriend-keeping. The best partner and the best beekeeper I could ever wish for. And to Miriam for bringing over the hive. I won the sibling lottery when I got you as my big sister. Thank you to some worker bees who buzzed on by when I needed them and helped me see this project through to its conclusion: Amabel Kylee Siorghlas, a patient editor bee from Vermont who stuck by me through the endless drafts; Daphne Gray-Grant, a motivator bee who helped me understand that a hive consists of about 40,000 bees and a book consists of about 80,000 words; her sister Jennifer, who always encouraged me to write; Naomi Pauls, an extraordinary proofreader and spelling bee; and Taryn Boyd, the queen bee of publishing, who took a chance on this useless drone who barged into her hive one day carrying a jar of honey and an early version of this book. A special thanks to my beta-copy readers: David Kincaid, Brian Antonson, Julie Prescott, Diane Zell, Mataya Varsek, Miriam Soet, and Jeannie Page. And thanks to one very knowledgeable beta reader who understands the fine line between humour and accuracy: Axel Krause, for his input and gentle push-back on that line. And finally, thanks to Rick Hansen—friend, mentor, hero, and lover of honey. The end is just the bee-ginning.